T0235349

Management and Industrial Engineering

Series editor

J. Paulo Davim, Aveiro, Portugal

More information about this series at http://www.springer.com/series/11690

Margarita Juárez-Nájera

Exploring Sustainable Behavior Structure in Higher Education

A Socio-Psychology Confirmatory Approach

Margarita Juárez-Nájera
Department of Energy
Universidad Autónoma Metropolitana
Mexico
Mexico

Management and Industrial Engineering
ISBN 978-3-319-38672-0   ISBN 978-3-319-19393-9   (eBook)
DOI 10.1007/978-3-319-19393-9

Springer Cham Heidelberg New York Dordrecht London

Printed on acid-free paper

Springer International Publishing AG Switzerland is part of Springer Science+Business Media
(www.springer.com)

Social processes are the ways in which our thoughts, feelings, and actions are influenced by the people around us, the groups to which we belong, our personal relationships, the teachings of our parents and culture, and the pressures we experience from others. Cognitive processes are the ways in which our memories, perceptions, thoughts, emotions, and motives guide our understanding of the world and our actions. Social and cognitive processes affect every aspect of our life, because the content of our thoughts, the goals towards which we strive, and the feelings we have about people and activities—all the ways we act and react in the social world—are based on what we believe the world is like

Eliot R. Smith and Diane M. Mackie

Not ideas, but material and ideal interest, directly govern men's actions

Max Weber

Man is naturally good

Rousseau

For Mariana

Preface

In 1998, I participated in the First Demonstrative Project of the Mexican Cleaner Production Center, whose goals included reducing contamination at the source, increasing businesses' economic gains, improving worker safety, optimizing productive processes, and incorporating more efficient technologies. During this project, and later in reviewing the results, I was continually dismayed by the lack of interest and involvement of a majority of the project's six participant entrepreneurs of the electroplating industry.

The project was supported by the United Nations Organization for Industrial Development (UNIDO) in collaboration with one of the most renowned technical higher educational institutions (HEI) in Mexico City, the *Instituto Politécnico Nacional*, and fully financed by the US Agency for International Development (USAID). Consequently, the only requirement for members of small- and medium-sized companies was a willingness to work with national and international electroplating cleaner production (CP) specialists, with whom they would document and implement environmentally and economically sound CP options.

Companies were selected based on their representativeness of the electroplating industry and according to number of employees, production volume, type of process, and financial capacity for investing in change. They were selected by well-known members of the Electroplating Industrial Association's technical committee. Prior to initiating, company leaders participated in a course on the scope, methodology, and expected outcomes of the project.

It seemed like a perfectly well-planned project, with its history of success in other countries, availability of financial and technical resources, and willingness of company leaders to participate. However, during the initial phase aimed at implementing suggested changes, two companies made no changes, and two others made only minor modifications. Only two companies made all suggested changes and even more, and achieved greater savings than expected.

I continually asked myself, "What happened?" "How is it possible that a majority of the companies had such a lack of interest in a project with free consulting and committed company leaders confident that they would financially benefit from CP changes?"

What impeded company leaders from implementing changes to their processes if they knew in return they would not only get their money back and increase savings, but also minimize environmental impacts and improve company processes and worker health and safety? What motivated them to participate in the CP project in the first place? Did company leaders initially act on behalf of their own interest, hoping for personal profit? Or did they act for the sake of others—the environment, workers, and the industrial association—encouraged by an altruistic spirit?

While these questions remain, I currently raise them with a deeper understanding as a result of decades as a professor in Higher Educational Institutions as well as my doctoral research. Furthermore, my research led me to apply the same questions to HEI: What drives decision makers' efforts in HEI? Do similarities exist between the behavior of individuals in higher education and that of company leaders? What factors determine the behavior of decision makers in HEI? Do decision makers foster the concept of sustainability in their activities, particularly when these activities take into account long-term implications for the institution and for social and cultural aspects of society?

In recent decades, higher educational institutions (HEI) have increasingly been forced to create, disseminate, and apply knowledge as a private property instead of a shared social construct or public good. This changing vision has sidetracked governments from their responsibility the principal providers of education, and, to some extent, HEI are more interested in obtaining profits than in resolving long-term problems such as environmental and social issues.

During the second year of my Ph.D. program, I realized the importance of social behavior in catalyzing and guiding decisions to implement change within organizations. During that period, I read a book on environmental policy and technological innovation titled, "Why Do Firms Adopt or Reject New Technologies?" by Carlos Montalvo-Corral. This book helped me begin to understand diverse aspects of individuals' resistance to change and gave me insight into some reasons for the lack, or slowness, of change with regard to environmental protection and implementation of CP approaches in industrialized and industrializing nations. Furthermore, I discovered theoretical frameworks, which could help me, to identify and work with the principal factors guiding human behavior in relation to resistance to change.

In addition, my awareness of the growing importance of education for sustainable development led me to the conclusion that education should be adapted to local contexts in order to attend to global problems. For the past 31 years, my professional life has been linked to environmental protection, both as a university professor and as an industrial consultant. Therefore, I have followed the evolution of the environmental education movement as well as new approaches such as that proposed by UNESCO, 2005, the Decade of Education for Sustainable Development (DESD) initiative.

The DESD initiative, and its recent edition from 2015 by Wals "Shaping the Education of the Future," has stressed the importance of influencing education at all levels to improve human life for present and future generations and to influence the general public to be more responsible for SD. Along with national and international pressure to bring about change in HEI is helping to involve faculty, administrators, other staff members, students, and alumni as agents of change. It is essential for academic leaders and other decision makers to increasingly support new ways to foster SD in education, research, outreach, and campus management.

It is urgent that the many decision makers of HEI in nations with varied cultural and economic structures become more aware of attitudes, policies, procedures, and practices which must be modified in order to help ensure that HEI truly foster SD. All those involved must work together to ensure that HEI faculty possess the knowledge and tools to educate present and future generations of students, and to ensure that decision makers become responsible in a rapidly changing world which is currently heading in unsustainable directions.

Organization of the Text

Chapter 1 provides an introduction to historical backgrounds, features, and underlying principles of ESD, as well as a definition of sustainable behavior in order to explore the main characteristics of ESD for present and future generations. Furthermore, the research goals are presented in this chapter.

Subsequently, the manuscript is divided into two parts according to the main research topics: personal factors in social psychology and areas of change. Part I is devoted to personality factors in social psychology and Part II to spheres of human intervention.

Part I explores people's motivations for acting in favor of the common good, as mentioned in the Decade of Education for Sustainable Development initiative: environmental conservation and protection, human rights, social security, gender equity, poverty reduction, health promotion, intercultural understanding and peace, sustainable consumption and production, and rural transformation. Also, it explores theoretical approaches suitable for devising a model for sustainable behavior and ways in which this model may be operationalized, tested, and validated.

Part I includes three chapters. Chapter 2 presents the theoretical framework, information processing approaches, which are part of cognitive theory, and some sociopsychological theories for determining factors of behavioral change. Chapter 3 presents the research method used to answer the research questions. Additionally, a model to determine sustainable behavior is proposed. This chapter also describes the methodology for applying and testing the sustainable-behavior model developed at five higher educational institutions in four countries with greatly different cultures and socioeconomic structures. Chapter 4 shows a statistical description of the specific factors of sustainable behavior; exploratory and confirmatory outcomes of the proposed model are discussed.

Part II explains the principles underlying education for sustainability in the UNESCO mandate of the Decade of Education for Sustainable Development. These principles include those areas of intervention in which people's beliefs may be modified in the long-term without coercion; factors which must be taken into account in order to achieve self-fulfilled citizens who are critical thinkers, equitable, fair, and responsible with respect to their environment, others, and themselves; and those activities which may be integrated into teaching, research, outreach, and campus managing within HEI in order to develop a way of life which foments education for sustainability. Chapter 5 points out to the differences between human needs and desires, and ways in which citizens may achieve self-fulfillment. Also, education and community management are described as two areas in which human behavior may be changed in the long-term without coercion.

Chapter 6 includes additional findings and comments on the scientific and practical value of the model developed, and a brief political reflection on these results.

Appendices A and B include a complete list of universal values and personal intelligences. Appendix C shows the English version of the questionnaire used at HEI when the original questionnaires (available by request to the author) were applied in Spanish, French, and German. Appendix D briefly explains multivariate statistical techniques used. Finally, the bibliography is presented at the end of each chapter.

Acknowledgments

I would like to end by acknowledging those who made this research possible. In chronological order according to the development of this study: Eduardo Campero allocated economic resources enabling me to carry out my doctoral studies without worrying about financial support; Don Huisingh and Leo Baas challenged, enthused, and invited me to participate in the International Off-Campus Doctoral Program at Erasmus University in Rotterdam. They have continually provided me with support in the planning and development of my doctoral dissertation. Additionally, other staff members at the Social Sciences Faculty—Nigel Roome, Jacqueline Cramer, and my very supportive Dutch advisor Wim Hafkamp—by commenting on my initial and final ideas.

Subsequently, I was pleased to meet and explore ideas with Carlos Montalvo-Corral, who inspired me to focus my research topic on his extensive study of social behavior and the application of environmental innovation. I deeply appreciate his willingness to guide me. I also wish to express gratitude to my local advisor Juan Rivera for his intense support in familiarizing me with the systemic vision and assisting me during the entire process of my doctoral work.

I am grateful for the opportunities I have had to discuss my doubts and progress with my colleague Hans Dieleman who, with his broad professional experience as a social researcher, grasped the importance of this research and offered many clear recommendations regarding the development of this study.

I am sincerely grateful for the invaluable support I obtained through cyberspace from pro-environmental psychologists Victor Corral-Verdugo and Florian Kaiser, without whom it would not have been possible for me to define the scope of my research. They provided me insight and information and helped me to clarify my vision, purpose, and direction.

I am indebted to Florian Wendelspiess for his generous support to run the *Mplus* software on the entire data set and to Mario Gonzalez-Espinosa for his kind and helpful suggestions of statistical tests at moments when I felt extremely pressured to carry out these tests. Thanks respectively to Manuel A. Gonzalez, Marcos Bustos, and Estelle Morin for opportune information on cognitive psychology, pro-environmental behavior and Maslow's hierarchy of needs.

I am sincerely grateful for Marco Reickmann's, Florian Wendelspiess', and Mariana Castellanos' extraordinary collaboration and invaluable willingness to encourage volunteers at *Leuphana Universität Lüneburg* in Germany, *Université de Genève* in Switzerland, and *Université de Montréal* and *Université de Québec à Montréal* in Canada, to respond by Internet to the questionnaire that has served as a contrast case for this research. Thanks also to Gerd Michelsen, Maik Adomssent, Horst Rode, and Clemens Mader for his institutional support at LUL.

Last but not least, thanks to my own institution *Universidad Autónoma Metropolitana, Azcapotzalco* (UAMA), for financial support during my doctoral studies and leaves of absence during my research at Rotterdam, Netherlands, in early 2009 and at Lüneburg, Germany, in spring 2013. UAMA provided me with the opportunity to conclude my doctoral study and finish this research. I claim full responsibility for its content.

Contents

List of Figures

List of Tables

Abbreviations

AC	Awareness of consequences
All-HEI	All members of Higher Educational Institutions
AR	Ascription of responsibility
CFA	Confirmatory factor analysis
DESD	Decade of Education for Sustainable Development
EE	Environmental education
ESD	Education for Sustainable Development
HEI	Higher Educational Institutions
HSELU	Higher Socio-Economic Level University
LSELU	Lower Socio-Economic Level University
LULIfUK	Leuphana Universität Lüneburg, Institut für Umweltkommunication, Germany
NAT	Norm-activation theory
PCA	Principal component analysis
PEB	Pro-environmental behavior
PI	Personal intelligences
TCHA	Theory of cultural-historical activity or Vigotsky's School
TMI	Theory of multiple intelligences
TPB	Theory of planned behavior
UAMA	Universidad Autónoma Metropolitana at Azcapotzalco, Mexico
UdeG	Université de Genève, Switzerland
UdeM	Université de Montréal, Canada
UNDP	United Nations of Development Program
UNEP	United Nations of Environmental Program
UNESCO	United Nations Educational, Scientific, and Cultural Organization
UNIDO	United Nations Organization for Industrial Development
UQAM	Université de Québec à Montréal, Canada
USAID	US Agency for International Development
UV	Universal values
VBNT	Value-belief-norm theory

Abstract

In this study, for the first time, the emergent concept of sustainable behavior (SB) by building a social-psychology model based on valid conceptual frameworks which are guided by principles of Education for Sustainable Development is elucidated; also to explore those spaces where beliefs and human behaviors may be modified. The SB model attempts to map those factors that may influence in an altruistic manner sustainable behaviour (SB) of students, faculty, and administrators in public higher education institutions (HEI) with very different economic and social characteristics. This model focuses on values and moral norms anchored in individuals instead of self-interest; also it is intended to compensate for deficiencies in explaining variances in models in favor of the environment. In order to test the SB model verifying the reliability and degree of association among latent variables considered—universal values, ascription of responsibility, awareness of consequences, and personal intelligences,—two statistical procedures were applied: principal component analysis explanatorily reveals a general pattern for the latent variables which underlie behavior for sustainability across HEI participants, and confirmatory factor analysis exposes evidence in the latent structure of a second order SB construct to understand the effect of their determinants. In order to develop critical, fair, responsible, self-actualizing citizens, this study considers two areas of human intervention for changing behavior in the long run without coercion: education and community management. It also proposes four methods as alternative forms of learning and ways of strengthening group change—play, art, group psychotherapy, and personnel management.

Chapter 1
Introduction

Due to a lack of critical considerations related to dignity, human rights, equity, care for the environment, and sustainable development, along with human diversity, inclusiveness, participation, and sufficiency for all; as well as major concerns that have demanded global attention such as HIV/AIDS, migration, climate change, and urban sprawl which reflect social, political, economic, and environmental challenges facing humanity and the planet, the UNESCO (2005) has promoted the Decade of Education for Sustainable Development (DESD), which, implemented in March 2005, emphasizes the importance of quality basic education and stresses that education must provide specific skills such as learning to know, learning to live together, learning to do, and learning to be (Delors et al. 1998).

The DESD and their 2009 and 2012 monitoring and evaluation reports are the most recent steps of a series of international resolutions organized by the United Nations. Efforts in education for sustainable development (ESD) may be traced back several Decades. The first part of this chapter explores the history of current sustainability efforts in development and education. Next, the key characteristics of ESD and a definition of sustainable behavior are presented. Finally, the research goal is proposed.

1.1 Background of Education for Sustainable Development

Sustainable development has its roots in the environment movement. Many important events have addressed sustainable development, including the 1972 World Summit on Human Environment held in Stockholm. Since then, numerous environmental protection agencies have been established, as well as the United Nations Environmental Program. While new programs studied social and economic aspects to some extent, greater priority was given to ecological incidents due to increasing uncontrolled development (UNEP 1972). Many nations realized that such generalized increased environmental degradation now required not only national approaches and solutions, but also international attention and collaboration.

In 1975, UNESCO, with the Belgrade Charter (UNESCO 1975), established a framework for environmental education (EE) to take into account environmental

© Springer International Publishing Switzerland 2015
M. Juárez-Nájera, *Exploring Sustainable Behavior Structure in Higher Education*,
Management and Industrial Engineering, DOI 10.1007/978-3-319-19393-9_1

protection mandates of the Stockholm World Summit (Orellana and Fauteux 1998). Such EE included the fundamental elements of the concept of sustainability:

- Formulation of basic concepts such as quality of life and human happiness, according to each particular culture;
- Re-formulation of the concept of development to focus on the satisfaction of needs and pursuits of all world citizens on the basis of social equality, justice, societal pluralism, and equilibrium between humans and the environment;
- A new universal economic order based on equality, absence of exploitation, peace, and disarmament;
- Addressing environmental and social problems on a global scale;
- Taking into account future generations;
- Change in value systems, life attitudes, and in relationships between humans and nature and among humans (Flogaitis 1998).

During the Tbilisi Conference in Russia in 1978, the UNESCO reaffirmed the guiding principles of EE: awareness, knowledge, attitudes, skills, and participation (Marcinkowski 2009) to include environmental, social, moral, economic, political, and cultural dimensions (Orellana and Fauteux 1998), thus reaffirming principles such as economic, political, and ecological inter-dependence; the relationships between economy, development, and environment; local and global perspectives; social and ecological responsibility; and solidarity among peoples and consideration for future generations (Flogaitis 1998; Hungerford 2009; Sauvé 1996).

All these ideas and goals place EE in the context of a movement of radical social, economic, and political change and educational reform with a global, interdisciplinary, problem-solving approach, values clarification and integration, critical thinking, experiential learning, and connection between schools and the broader community (Flogaitis 1998; Hungerford 2009).

Within 10 years after the Stockholm summit, the world community began to realize that treating environmental concerns in isolation of development needs was not benefiting either the environment or people (UNESCO 2005). Hence, by the mid-1980s, the United Nations launched a search for a broader strategy which could address both social and environmental needs. In 1987 with the Brundtland Commission Report, "Our Common Future" (UN 1987), sustainable development was endorsed as an overarching framework for future development policy at all governmental levels.

In the 1992 World Summit on Environment and Development in Rio de Janeiro, EE was inextricably linked with sustainability (Orr 1992). As such, EE was recognized as a fundamental tool for achieving environmental goals. However, several studies (Bravo-Mercado 2005; Dieleman and Juarez-Najera 2007; Eilam and Trop 2010; Eschenhagen 2007; Flogaitis 1998; González-Gaudeano 1997; UNEP 2003) show that the majority of EE programs have traditional environmental studies, naturalist approach. Such EE promotes concepts and tools which tend toward technocratic solutions, with no connection to the root cause of environmental and social problems. This approach involves an absence of questioning and critical consideration of political, social, and economic issues relevant to environmental issues and therefore cannot play an essential role in fomenting changes required to achieve sustainable development (Eilam and Trop 2010; Flogaitis 1998; Stevenson 2007).

This misguided EE approach stems from the social and educational *status quo*, including inadequate teacher training, difficulties in the design and practice of interdisciplinary approaches, and isolation of schools from their communities. EE, as part of the environmental movement, touches on social, political, and ideological confrontations arising from environmental issues, thus requiring a variety of solutions and priorities. Since a technical, managerial approach governs society in general, and environmental issues in particular, this reflects the type of EE ultimately put forward (Flogaitis 1998). The challenge of sustainability demands reconsideration and reorientation of the conformist approaches to EE and reinforcement of a critical, participatory EE toward social, political, and educational changes.

The concept of education for sustainability was initially included in Chap. 36 of UNESCO's 1992 Agenda 21: "Promoting Education, Public Awareness, and Training." In addition, education as a strategy to promote and implement environmental change was embedded in each of the 40 chapters of the Agenda (Keating 1995) and in each of the post-Rio United Nations Conferences in the 1990s.

The World Summit on Higher Education, which focused on higher education, was organized at Paris in 1998. This conference reinforced the mission of higher education "to educate, train, carry out research and, particularly, contribute to sustainable development and improve society as a whole." The mission states that education in general, and higher education in particular, is the fundamental pillar of human rights, democracy, sustainable development, and peace (UNESCO 1998).

The 2002 Johannesburg World Summit on Sustainable Development reaffirmed the importance of sustainable development as a basis for overcoming poverty and improving quality of life worldwide, especially in the so-called developing world. As a follow-up to "Johannesburg," in December 2002, the United Nations General Assembly adopted the resolution "DESD," proposed by Japan and cosponsored by 46 countries. UNESCO ratified the resolution in April, 2003.

The DESD resolution is emphasizing that education for all is a vital condition for sustainable development. The crucial message of the "Decade" to the world is that "education is the primary agent of transformation toward sustainable development" (UNESCO 2005). Education has the capability of increasing people's capacity to transform their visions for society into reality. Education not only provides scientific and technical skills, but also provides the motivation, justification, and social support for pursuing and applying these skills (Juarez-Nájera et al. 2006a).

The DESD is a transformational undertaking because it entails that ESD focuses on underlying principles and values conveyed through education. As such, ESD is concerned with the content and purpose of education, and, more broadly, with all types of learning. ESD is a challenge for all forms of education and includes pedagogical processes, validation of knowledge, and the functioning of educational institutions.

In 2007, UNESCO adopted a resolution on ESD that 'recognized that further substantial initiatives have to be taken by Member States and by UNESCO in order to reorient teaching and learning toward SD worldwide'. These goals, as well as UNESCO's role in supporting their achievement, were reiterated at the mid-DESD conference held in Bonn, Germany (UNESCO 2009; Wals 2009). The Bonn Declaration also gave the world an action plan for ESD and provided concrete steps for implementing the remainder of the Decade.

The monitoring and evaluation of the DESD occurs within the Global Monitoring and Evaluation Framework (GMEF) designed by the DESD Monitoring and Evaluation Expert Group (MEEG). After its first meeting in 2007, the MEEG recommended that UNESCO (2007) publish three DESD implementation reports during the life of the Decade:

in 2009: focusing on the context and structures of work on ESD in Member States;
in 2011: focusing on processes and learning initiatives related to ESD; and
in 2015: focusing on impacts and outcomes of the DESD.

That is, during the second half of the DESD, the focus has shifted toward realizing visible results.

As part of the current monitoring and evaluating report of the DESD's progress, the GMEF has proposed a Phase I of revision, completed at the mid-DESD point, focused on structures, provisions, and policies put in place during the first half of the DESD to support the development of ESD around the globe (UNESCO 2012).

After the Rio Earth Summit+20, at a time when the challenge of SD is as great as ever (Worldwatch 2012), a second report appears and represents the outcome Phase II of the GMEF. It focuses on the learning processes taking place in the various contexts of education, teaching and learning both in the public and private sectors and in the intersections between them as a result of people engaging in ESD.

The report states that there is increased recognition that this challenge cannot be solved only through technological advances, legislative measures, and new policy frameworks (UNEP 2011). While such responses are necessary, they will need to be accompanied by changes in mind-sets, values, and lifestyles, as well as a strengthening of people's capacities to bring about change. This recognition explains the key role many governments, NGOs, UN Agencies and indeed, companies are allocating to learning and capacity-building as they search for solutions to interrelated sustainability challenges such as climate change, disaster risk management, biodiversity loss, sustainable production, and consumption (UNESCO 2012).

1.2 Key Characteristics of Education for Sustainable Development

The UNESCO (2005) initiative emphasizes that no universal models exist; rather, education depends on local contexts, priorities, and approaches. This initiative recommends that goals, emphases, and processes must therefore be locally defined to meet local environmental, social, and economic conditions in culturally appropriate manners. It also states that ESD is essential to all nations, regardless of their greatly varied cultures and socioeconomic structures.

In order to achieve ESD, the UNESCO (2005) identifies four principal ways in which education may support a sustainable future: (1) improving access to quality basic education, (2) reorienting existing educational programs, (3) developing public understanding and awareness of sustainability, and (4) providing training in sustainability issues.

The UNESCO resolution points out essential characteristics of ESD, which may be implemented in a variety of culturally appropriate ways (UNESCO 2005, 2012). The list below presents some of the features of such education along with an explanation for each.

1. ESD is based on principles and values which underlie sustainable development, including the tenet that education is a human right.
2. ESD deals with economic, social, and cultural sustainability (Elkington 1998), that is, a just, equitable, and peaceable world in which social tolerance and gender equity is practiced and people care about the environment and thus contribute to natural resource conservation, inter-generational equity, and poverty alleviation (UNESCO 2005).
3. ESD promotes lifelong learning (UNESCO 2005), which, broadly understood, describes a process in which individuals, with the help of others, diagnose their needs for learning and education, formulate their goals, identify their resources, select and implement their strategies, and evaluate their educational outcomes (Castrejón and Ángeles 1974; Commission of the European Communities 2000; Ramnarayan and Shyamala 2005).
4. ESD is locally relevant and culturally appropriate, based on local needs, perceptions, and conditions, and acknowledges that fulfilling local needs often has international effects and consequences (UNESCO 2005).
5. ESD engages formal, non-formal, and informal education (UNESCO 2005). Formal education takes place within educational institutions and leads to the acquisition of grades and diplomas. Non-formal learning occurs in a formal learning environment, but is not officially recognized within a curriculum. Informal learning occurs through experiences in daily situations. Both types of education are compatible with formal education and normally do not lead to certificates. Informal learning is unintentional and the learner is often unaware of the process. Nowadays, formal learning dominates political thought, establishing the manner in which education is provided. Non-formal and informal learning are typically undervalued (Castrejón and Ángeles 1974).
6. ESD must be adapted to the evolving nature of the concept of sustainability, not only to environmental, social and economic areas, but the sustainability concept must include seven dimensions, as explained by Morin (2001) and Morin et al. (2002) and summarized by Dieleman (2000): (1) thematic dimension: ecology, economy, and social equity; (2) spatial dimension: north–south dialogue; (3) temporal dimension: relevant to the present and preparing individuals for the future; (4) spiritual dimension: a sense of belonging to the whole;

(5) institutional dimension: social change; (6) esthetic dimension: beauty, use of materials; and (7) knowledge dimension: systemic thinking.

7. ESD takes into consideration global problems and national priorities and adjusts the syllabus to these unique conditions (UNESCO 2005).
8. ESD builds civic capacity for community-based decision making, social tolerance, environmental stewardship, and high quality of life; ESD also promotes competency of the learner as an individual, a family member, a community member, and a global citizen (UNESCO 2005).
9. ESD is interdisciplinary, building knowledge, life skills, perspectives, attitudes, and values (UNESCO 2005).
10. ESD uses a variety of pedagogical techniques which promote participatory learning and higher-order thinking, makes use of playful learning, and provides tools to transform actual societies into more sustainable societies (UNESCO 2005).
11. ESD is measurable (UNESCO 2005).
12. ESD focuses on performance and seeks collective success, development, well-being and quality of life; ESD is flexible and liberating (UNESCO 2005).
13. ESD, based on humanistic principles, is capable of educating people to become more humanistic, thus learning to live together. Such positive interaction leads to strengthening appreciation for human dignity, the desire for social well-being, support of ideals such as fraternity, equal rights for all, abolition of privilege according to race, religion, gender, or other individual qualities, and the implementation of international solidarity and sustainability (Benavides 1998).

ESD is fundamentally about values (UNESCO 2004), with the fundamental pillar being consideration and respect for others, including present and future generations, respect for cultural and societal difference and diversity, for the environment, and for planetary resources. Education enables us to understand ourselves and others and our links with the broader natural and social context (Benavides 1998), and this understanding serves as a basis for building respect.

Table 1.1 shows ESD principles and characteristics. ESD is based on a holistic vision and an interdisciplinary, values driven, critical thinking approach focused on problem solving in local, participatory decision making, taking advantage of pedagogical, recreational, and artistic methods. Education for sustainability must enable students to understand the complexity of global environmental, social, and cultural settings. ESD proposes sustainable alternatives to current practices. Students must understand that in order to attend to the current situation, they must develop a critical, responsible, and participatory attitude based on sustainability and that the analysis and solutions are inter- and transdisciplinary.

The DESD is highly ambitious as has sought to affect multiple levels of governance and to engage multiple (including marginalized) stakeholders. There is no doubt that people are engaged worldwide in ESD in a variety of ways as demonstrated in the two-year progress report (UNESCO 2007), and in the first and second global monitoring reports (UNESCO 2009, 2012). The DESD implementation has made considerable progress since its international launch in March 2005.

Table 1.1 Principles of education for sustainability (based on UNESCO 2005, 2012)

Characteristics	Principle
Learning for sustainable development embedded in the whole curriculum, research, outreach and management campus programs, not as a separate subject	Interdisciplinary and based on systems thinking
Sharing the values and principles underpinning sustainable development	Values driven
Leading to confidence in addressing the dilemmas and challenges of sustainable development	Critical thinking and problem solving
Art, debate, drama, playful experiences, different pedagogies, etc. which model the learning processes	Multimethods
Learners participate in decisions on how they are to learn	Participatory decision making
Addressing local as well as global issues, and using the languages which learners most commonly use	Locally relevant, effective, and contextual

Twenty years after Chap. 36 of Agenda 21, the "Shaping the Education of the Future" (UNESCO 2012; Wals 2014) report (the second DESD global monitoring report) provides a good opportunity to see to what extent the practice learning processes and multistakeholder interactions that engage in such profound change, often times involving the development of alternative values, are still scarce around the globe. Still there are strong indications that people within and outside of ESD require new forms of learning, professional development, competencies, and monitoring, evaluation and assessment.

As the DESD approaches 2014—its final year—supporting and further developing ESD as a catalyst for reorienting education, teaching, learning, and professional development toward more holistic, integrative, and critical ways of addressing sustainability challenges is paramount (UNESCO 2012). This will require strengthening capacity-building. A whole-system approach that affects all actors in a school system or a production chain seems the most likely to ensure such capacity-building and competence development. This way ESD may help to contribute to SD.

1.3 Definition of Sustainable Behavior

The idea for this study of sustainable behavior arises from two viewpoints: environmental psychology, and sustainability as an evolving concept. Environmental psychology explores the interaction between people and their physical setting (Corral-Verdugo and Pinheiro 2004), or in other terms, the relationship between people (human well-being) and the broader environment (socio-physical context) (Corral-Verdugo 2001). The concept of sustainability has its roots in the "green" movement of the United States and Europe since the late 1960s. During this period, Western society has become more conscious of living in harmony with nature, the limits to natural resources, and the worsening environmental problems (Bonnes and Bonaiuto 2002).

All this has caused a change in world views regarding human exception, by which the human being is conceived as a special organism—an exception among animal species. Animals basically depend on their instincts in order to survive. Humans, on the other hand, have markedly different learning mechanisms, act with deliberation, and are capable of dominating other organisms (Corral-Verdugo 2001; Corral-Verdugo and Pinheiro 2004). This world vision has shifted toward a new environmental paradigm (Dunlap and van Liere 1978; Dunlap et al. 2000), which holds that humans are part of the natural world and subject to rules of nature, and are governed by the inter-dependence of species. Earlier behavioral theoretical approaches such as Skinner's contingency model stated that conditions that exist when a response is followed by a reinforcement action enable a range of environment–behavior relationships to satisfy a contingency (Corral-Verdugo 2001). The newer cognitivist model aims to study the information determinants of thought processes and related events. That is, behavior is influenced by the information an organism stores in the brain and the brain's information processing systems (Medin et al. 2005; Von Eckardt 1996). Finally, this new paradigm moves from a disciplinary approach toward inter-, multi-, and transdisciplinary approach, which transcends disciplines to address any problem (Nicolescu 2008).

Human behavior in general, according to Corral-Verdugo and Pinheiro (2004), is composed of many facets. For example, those aspects dealing with problem resolution are considered to be related to competence or performance; those dealing with choice or preference are called motives or attitudes; those facets related to objects or events according to cultural norms are called beliefs; idiosyncrasies, or aspects which reflect the individual's peculiarities, are considered to be related to personality. All facets are involved in intentional and irrational actions.

According to the Merriam-Webster online dictionary (see: http://www.m-w.com/dictionary/behavior), the word "behavior" has three definitions: (1) the manner of conducting oneself, (2) anything that an organism does involving action and response to stimulation, and (3) the response of an individual, group, or species to their environment. For this study, the third definition is most appropriate, since the first is not related to the environment and the second responds to a theoretical framework which has provided many ideas toward the formulation of studies of pro-environmental conduct, but which few authors currently consider.

Several authors (Corral-Verdugo 2001; Kaiser 1998; Kantor 1967) consider behavior to be the interaction between organisms and objects. Specifically, pro-environmental behavior (PEB) is defined as actions contributing to environmental conservation, or human activity intended to protect natural resources or at least reduce environmental deterioration.

These definitions include a deliberate component, or intentionality, and expect a result. In conclusion, sustainable behavior has three main characteristics: (1) it is an outcome or result; (2) it is effective, and (3) it is complex.

PEB is effective because it consists of actions that generate visible changes in the environment. PEB is also a product or outcome, since it is a response to requirements or a solution to problems. This means that PEB must be analyzed as competencies, that is, as effective responses facing demands for environmental protection, and also as behavior, that is, as deliberate effective responses taking responsibility for environmental protection. These demands may be individual attitudes or motives, or social norms. Therefore, the study of beliefs and attitudes is indispensable. PEB has a high level of complexity because it allows us to anticipate a situation and plan ahead in order to achieve effective results. This reinforces the need to study norms and values that an individual establishes as a framework for carrying out pro-environmental actions on a continual basis.

Considering the previous characteristics, PEB can be defined as a "set of deliberate and effective actions which respond to social and individual requirements for protecting the environment," and sustainable behavior as an intentional behavior aimed at protecting the environment and encouraging human well-being and security.

According to the aforementioned, and adapted from the definition by Corral-Verdugo and Pinheiro (2004), sustainable behavior is "a set of effective, deliberate, and anticipated actions aimed at accepting responsibility for prevention, conservation and preservation of physical and cultural resources. These resources include integrity of animal and plant species, as well as individual and social well-being, and safety of present and future human generations." This extensive definition provides a point of reference for determining sustainable human behavior in this study.

Three main differences exist between this definition and that considered by Corral-Verdugo and Pinheiro (2004):

1. This definition considers responsibility, that is, the capacity for responding or acting instead of competing.
2. It addresses prevention and conservation, not only preservation.
3. It includes individual and societal material safety.

These modifications to the definition, according to the author of this study, make the definition more complete. First, they are directed toward an effective disposition toward problem solving with the taking of responsibility by individuals; that is, individuals are willing to resolve problems through actions or behavior. These behaviors are considered in the characteristics of ESD in order to educate citizens capable of responding to future demands (Juárez-Nájera 2007).

Second, this definition considers not only the conservation and preservation of physical environment, but also prevention. These aspects have been controversial topics since the mainstream environmental protection movement began to address environmental deterioration. Preservation consists of covering up damage or danger, but can sometimes be an essential, if not sufficient, element of conservation. Conservation, on the other hand, refers to maintaining the environment in its original state. Both aspects are important. However, the principle of prevention has never been explicit. This principle draws on knowing, preparing, and taking action in order to avoid environmental deterioration. This definition takes into account conservation, preservation, and prevention.

Third, the original definition considers only human well-being, not future security of natural resources. Nevertheless, in order to assure long-term sustainability, according to Gardner and Stern (2002), the following must be accomplished:

1. Exponential human population growth must be halted.
2. Economic and material growth must be controlled, and such growth must be oriented toward qualitative development rather than physical expansion, and toward material sufficiency and security for all.
3. Profound changes must be made in core societal beliefs, values, and ethics concerning population growth, material growth, wealth, and well-being, as well as in basic conceptions of the relationship between humans and the rest of nature, acknowledging the complexity of global systems and humanity's inability to manage these systems solely for our own purposes.

1.4 Research Goal

This study explores the research question of identifying psychological factors related to social, cultural and personality features, which can influence sustainable behavior of individuals (student, faculty, staff, and administrator) in higher education institutions (HEI) in nations with greatly varying cultures and socioeconomic structures, as well as to present the areas where these individuals work, and in which higher education for sustainability is fostered.

United Nations international agencies are shifting their support toward ESD, rather than EE. For over 40 years, EE has focused mainly on the biological dimensions of issues, the role of human beliefs and values, and the roles played by the citizen, individually and collectively, in environmental issue resolution (Hungerford 2009) as separated from social and ethical problems. Although the founding documents of EE include elements of the concept of sustainability, the UNESCO (2005) resolution of the DESD placed EE program in a smaller division, whereas the ESD program was enlarged and given greater prominence because is much broader in scope than EE (Marcinkowski 2009).

The challenge is to devise ways to achieve socially desirable goals, such as the ones underlying the goals of ESD, for example to involve actor participation in decision making regarding local conditions, directed toward values and the development of critical thinking; they focus on problem solving, are based on methods such as playing games and art appreciation, and be holistic and interdisciplinary (UNESCO 2005, 2012).

The UNESCO (1998, 2004, 2005, 2009, 2012) has pointed out that education in general, and higher education in particular, is the cornerstone of human rights, democracy, sustainable development, and peace. In order to propose an alternative higher education, it is important to understand and identify ways in which behavior may be affected. Factors involved in achieving the determinants of human behavior toward a responsible citizenry who seek equality, justice, peace, and the public good should be reviewed.

To begin, it is important to define the meaning of sustainable behavior. In this study, it is considered to be the set of effective and deliberate actions directed toward conservation and/or preservation of physical and cultural resources, integrity of animal and plant species, and individual and social well-being and safety of present and future generations (Juárez-Nájera et al. 2010, see Sect. 1.3).

This definition leads us to a theoretical framework based on current psychological developments in cognitive psychology in order to explore people's changing attitudes. The cognitive science approach, developed in the 1960s, suggests that learning and cognition depend on individuals' cognitive information processing (Medin et al. 2005; Von Eckardt 1996).

This study is also based on social psychology, which is the scientific study of the social nature of humans (Jahoda 2007), or the effects of social and cognitive processes on the way individuals perceive, influence, and relate to others (Smith and Mackie 2007). That is, the reciprocal influences of the individual and his or her social context through the behavioral expression of that individual's thoughts and feelings within her or his society and culture considered both synchronically and diachronically (Manstead and Hewstone 1995).

Therefore, the model of sustainable behavior presented here addresses a range of contexts, from intrapersonal processes and interpersonal relations to inter-group behavior and societal analyses. That is, on the one hand, social processes are the ways in which our thoughts, feelings, and actions are influenced by the people around us, the groups to which we belong, our personal relationships, the teachings of our parents and culture, and the pressures we experience from others. On the other hand, cognitive processes are the ways in which our memories, perceptions, thoughts, emotions, and motives guide our understanding of the world and our actions. Social and cognitive processes affect every aspect of our life, because the content of our thoughts, the goals toward which we strive, and the feelings we have about people and activities—all the ways we act and react in the social world—are based on what we believe the world is like (Smith and Mackie 2007).

Within social psychology, two main conceptual frameworks explain human behavior: that which is based on self-interest and that based on altruism (Kaiser et al. 2005). Rousseau (2001), in his eighteenth century "Discourse on Inequality," states that humans act based on either egoism or selflessness, but regardless, human goodness should be fostered so that individuals continue to be humane (Neuhouser 2008, 2014). This study considers the conceptual framework of norm activation, based on moral norms grounded within individuals (Schwartz 1977; Schwartz and Boehnke 2004). Personal norms, if activated, are experienced among individuals as feelings of personal obligation, either denying or not denying the consequences of their behavioral choices regarding the welfare of others.

In order to propose a model to test behavior, specifically sustainable behavior, two models for modifying behavior toward pro-environmental action were studied: the meta-analysis by Hines et al. (1987), and the model proposed by Stern et al. (1999) based on motivational values and two personality traits of norm activation. Also, two of the seven types of human intelligence inter- and intrapersonal intelligence described by Gardner (2001) in his theory of multiple intelligences demonstrated in all cultures were added. These aspects were chosen due to the ease of demonstrating these qualities through a written test. These two skills were then analyzed through the five Corral-Verdugo and Pinheiro's (2004) and Corral-Verdugo's (2010) psychological dimensions, which are based on the notion of sustainability: effectiveness, deliberation, anticipation, solidarity, and austerity. Consequently, the model focuses on values and moral norms rather than on rational choice and self-interest, while allowing people to recognize such moral norms through latent variables such as values, ascription of responsibility, awareness of consequences, and personal skills, as ways of explaining their behavior.

In order to operationalize the proposed model, a 67-item questionnaire was developed based on the model of Stern et al. (1999). This model was updated to include the four types of motivational values proposed by Schwartz (1977, 1994), Schwartz and Bilsky (1987, 1990), Schwartz and Huismans (1995), and Schwartz and Boehnke (2004). These authors suggest that these four types of values are present in all humans worldwide. Two variables, the new ecological paradigm and cultural models, were omitted because they showed low reliability in previous studies (Kaiser et al. 2005). Also, two personality traits (ascription of responsibility and consciousness of consequences) were considered by asking about environmental topics. Twenty of the 72 items for emotional competencies of Boyatzis et al. (2002) were adapted through five psychological dimensions. Finally, 6 demographical items (gender, age, income as who own or rent a house or apartment, religious denomination, activity, and level of education) were added. The first and most obvious goal of the model was to improve the explanation of behavioral variance compared to previous models, as proposed by Corral-Verdugo (2001), Harland (2001), and Stern (2000).

The questionnaire was applied to 127 individuals in a Mexican HEI, and 85 in a Germany HEI, 19 in a Switzerland HEI, and 9 in French-speaking Canada HEI. The first is the *Universidad Autónoma Metropolitana, Azcapotzalco* (UAMA), which is located north of Mexico City and is a public university. The second university is the *Leuphana Universität Lüneburg, Institut für Umweltkommunication* (LULIfUK), a public university near Hamburg in the Federal Republic of Germany, honored with the UNESCO Chair in Higher ESD. The third is the *Université de Genève* in Switzerland, and the fourth is both the *Université de Montréal* and *Université de Québec à Montréal* in French-speaking Canada.

In order to validate the proposed model, two statistical methods were applied in the following order: principal component analysis and confirmatory factor analysis.

In order to identify the areas in which key individuals within HEI work and the ways in which ESD is fostered, two areas of intervention in which people may plausibly modify their beliefs without coercion for the long run were presented: education and community management (Gardner and Stern 2002). These areas include four HEI activities, teaching, research, outreach, and campus physical operations and may potentially make use of alternative learning methods and group projects, play, art, psychotherapy groups, and labor management as ways to foster behavior toward sustainability (Juárez-Nájera et al. 2006b, 2010).

Play is fun, relaxing, and holistic, and failure does not cause damage (Dieleman and Huisingh 2006); art inspires awe (Gordon and Gordon 2008), which also comes into play in the appreciation of nature, and is a necessary step in experiencing a desire to take care of the environment (Juárez-Nájera et al. 2006a, b). These are helpful tools, because achieving sustainability requires changes in pedagogy, and play and art provide holistic ways of learning about reality, while science, with its analytical rationality, when applied to grasp reality, cannot express desires, emotions, fears, lifestyles, identities, and intuitive notions (Dieleman 2007a, b, c). In the community management area, psychotherapy groups and labor management can foment self-esteem in order to achieve self-assured citizens (Maslow 2005) able to work for a better, more sustainable world.

References

Benavides, L. G. (1998). *Hacia Nuevos Paradigmas en Educación*. México: CIPAE.

Bonnes, M., & Bonaiuto, M. (2002). Environmental psychology: From spatial-physical environment to sustainable development. In R. Bechtel, & A. Churchman (Eds.), *Handbook of environmental psychology* (Chap. 3, pp. 28–54). NYC: Wiley.

Boyatzis, G., & Hay Acquisition Co. Inc. (2002). *Inventario de competencias emocionales*. HayGroup.

Bravo-Mercado, T. (2005). *El Cambio Ambiental de las Instituciones de Educación Superior: Avances y Retos*. En: Eduardo S. López-Hernández, Ma. Teresa Bravo Mercado, Édgar J. González Gaudiano (Coordinadores): *La profesionalización de los educadores ambientales hacia el desarrollo humano sustentable*. Serie Memorias Colección Biblioteca de la Educación Superior, México: ANUIES.

Castrejón, J., & Ángeles, O. (1974). *Educación Permanente. Principios y experiencias*. 1a. Edición. México: Fondo de Cultura Económica.

Commission of the European Communities. (2000). Commission Staff Working Paper. A Memorandum of Lifelong Learning. Brussels, Oct 30. SEC (2000) 1832.

Corral-Verdugo, V. (2001). *Comportamiento Proambiental. Una introducción al estudio de las conductas protectoras del ambiente*. Canarias: Editorial Resma.

Corral-Verdugo, V. (2010). *Psicología de la sustentabilidad: un análisis que nos hace pro ecológicos y pro sociales*. México: Trillas.

Corral-Verdugo, V., & Pinheiro, J. (2004). Aproximaciones al estudio de la conducta sustentable. *Medio Ambiente y Comportamiento Humano, 5*(1), 1–26.

Delors, J. et al. (1998). *Learning: The treasure within*. Report to UNESCO of the International Commission on Education for the Twenty-first Century. Geneva: UNESCO.

Dieleman, H. (2000). *Guidelines for a more effective training and education in environmental management*. Varese, Italy: EAEME.

Dieleman, H. (2007a). The competencies of artful doing and artful knowing in higher education for sustainability. In M. Adossment, A. Beringer, & M. Barth (Eds.), *World in transition—sustainability perspectives for higher education part II* (pp. 62–71). Germany: Vas.

Dieleman, H. (2007b). Science and Art in Society and Sustainability; about a carpenter, a hammer and a chisel. Article published in the webmagazine of Cultura, July 21, 2007, special issue on science and art in sustainability.

Dieleman, H. (2007c). Sustainability, art and reflexivity; why artist and designers may become key change agents in sustainability. In: Proceedings of the ESA conference: New Frontiers in Arts Sociology. Lüneburg and Hamburg, March 28–April 1, 2007.

Dieleman, H., & Huisingh, D. (2006). The potentials of games in learning and teaching about sustainable development. *Journal of Cleaner Production, 14*(9–11), 837–847.

Dieleman, H., & Juárez-Nájera, M. (2007). State of the art of higher education for sustainability in Mexico; analysis of 40 institutional environmental plans on 7 indicators. In M. Adossment, A. Beringer, & M. Barth (Eds.), *World in transition—sustainability perspectives for higher education (part IV* (pp. 206–215). Germany: Vas.

Dunlap, R. E., & van Liere, K. D. (1978). The new environmental paradigm. *Journal of Environmental Education, 9,* 10–19.

Dunlap, R. E., Van Liere, K. D, Mertig, A. G., & Jones, R. E. (2000). Measuring endorsement of the new environmental paradigm. A revised nep scale. *Journal of Social Issues, 56*(3), 425–442.

Eilam, E., & Trop, T. (2010). ESD pedagogy: A guide for the perplexed. *The Journal of Environmental Education, 42*(1), 43–64.

Elkington, J. (1998). *Cannibals with forks: The triple bottom line of 21st century business.* NYC: Capstone, 1997/New Society, 1998.

Eschenhagen, M. L. (2007). La educación ambiental superior en América Latina: una evaluación de la oferta de posgrados ambientales. *Argentina Theomai 16,* 1–21.

Flogaitis, E. (1998). *The contribution of environmental education in sustainability.* Greece: University of Athens.

Gardner, H. (2001). *Estructuras de la Mente. La teoría de las inteligencias múltiples.* Segunda Edición. México: Fondo de Cultura Económica (1993).

Gardner, O. T., & Stern, P. C. (2002). *Environmental problems and human behavior* (2nd ed.). NYC: Pearson Custom Publishing.

González-Gaudeano, E. (1997). *Educación Ambiental, historia y conceptos a veinte años de Tbilisi.* Mexico, D.F.: Sistemas Técnicos de Edición, S.A. de C.V.

Gordon, R., & Gordon, H. (2008). *Hobbema and Heidegger. On truth and beauty.* New York: Peter Lang.

Harland, P. (2001). *Pro-environmental behavior.* Doctor Thesis. Faculteit der Wiskunde en Natuurwetenschappen en die der Geneeskunde, Universiteit Leiden, the Netherlands.

Hines, J. M., Hungerford, H. R., & Tomera, A. N. (1987). Analysis and synthesis of research on responsible environmental behavior: A meta-analysis. *Journal of Environmental Education, 18* (2), 1–8.

Hungerford, H. R. (2009). Environmental education (EE) for the 21st century: Where have we been? Where are we now? Where are we headed? *The Journal of Environmental Education, 41* (1), 1–6.

Jahoda, G. (2007). *A history of social psychology. From the eighteenth-century enlightenment to the second world war.* Cambridge: Cambridge University Press.

Juárez-Nájera, M. (2007). Human tasks based on competence or on performance? A perspective toward higher education for sustainable development. In: Proceedings of the Second International Conference on Sustainable Development for Higher Education "World in Transition—Sustainability in Perspective". July 5–7 at Autonomous University of San Luis Potosi, Mexico.

Juárez-Nájera, M., Dieleman, H., & Turpin-Marion, S. (2006a). Sustainability in Mexican higher education, towards a new academic and professional culture. *Journal of Cleaner Production, 14*(9–11), 1028–1038.

Juárez-Nájera, M., Dieleman, H., & Turpin-Marion, S. (2006b). Games as tools for sustainability: The usage in education and in community outreach. In Proceedings of the 4th International Conference on environmental Management for Sustainable Universities. June 26–30, at Stevens Point, Wisconsin, USA.

Juárez-Nájera, M., Rivera-Martínez, J. G., & Hafkamp, W. A. (2010). An explorative socio-psychological model for determining sustainable behavior: Pilot study in German and Mexican Universities. *Journal of Cleaner Production, 18,* 686–694.

Kaiser, F. G. (1998). A general measure of ecological behavior. *Journal of Applied Social Psychology, 28*(5), 395–422.

Kaiser, F. G., Hübner, G., & Bogner, F. (2005). Contrasting the theory of planned behavior with the value-belief-norm model in explaining conservation behavior. *Journal of Applied Social Psychology, 35*(10), 2150–2170.

Kantor, J. R. (1967). *Interbehavioral psychology. A sample of scientific system construction.* Grainville, Ohio: The Principia Press. (1959).

Keating, M. (1995). *The earth summit's agenda for change: A plan language version of agenda 21 and other Rio agreements. The center for our common future,* (4th ed.). Geneve, Switzerland: Atar, S.A.

Manstead, A. S. R., & Hewstone, M. (Eds.). (1995). *The blackwell encyclopedia of social psychology.* UK: Blackwell Publishers.

Marcinkowski, T. J. (2009). Contemporary challenges and opportunities in environmental education: Where are we headed and what deserves our attention? *The Journal of Environmental Education, 41*(1), 34–54.

Maslow, A. (2005). *El management según Maslow (Maslow on Management).* Paidós: Una visión humanista para la empresa de hoy. Barcelona.

Medin, D. L., Ross, B. H., & Markman, A. B. (2005). *Cognitive psychology* (4th ed.). New York: Wiley.

Morin, E. (2001). *Los Siete Saberes Necesarios para la Educación del Futuro.* México: Librería El Correo de la UNESCO.

Morin, E., Ciurana, E. R., & Motta, R. D. (2002). *Educar en la Era Planetaria. El pensamiento complejo como método de aprendizaje en el error y la incertidumbre humana.* Salamanca: Universidad de Valladolid.

Neuhouser, F. (2008). *Rousseau's theodicy of self-love. evil, rationality, and the drive for recognition.* New York: Oxford University Press.

Neuhouser, F. (2014). *Rousseau's critique of inequality: Reconstructing the second discourse.* Cambridge: Cambridge University Press.

Nicolescu, B. (Ed.). (2008). *Transdisciplinarity—theory and practice.* New Jersey, USA: Hampton Press.

Orellana, I., & Fauteux, S. (1998). *Environmental education: Tracing the high points of its history.* Montreal: Université du Québec.

Orr, D. (1992). *Ecological literacy: Education and the transition to a postmodern world.* New York: State of New York Press.

Ramnarayan, K., & Shyamala, H. (2005). *Thoughts on self-directed learning in medical schools: Making students more responsible.* Seattle, USA: New Horizons for Learning.

Rousseau, J. J. (2001). *Discurso sobre el origen de la desigualdad entre los hombres.* Madrid: Diana.

Sauvé, L. (1996). Environmental education and sustainable development: a further appraisal. *Canadian Journal of Environmental Education, 1,* 7–33.

Schwartz, S. H. (1977). Normative influences on altruism. In L. Berkowitz (Ed.), *Advances in experimental social psychology* (Vol. 10, pp. 221–279). New York: Academic Press.

Schwartz, S. H. (1994). Are there universal aspects in the structure and contents of human values? *Journal of Social Issues, 50,* 19–45.

Schwartz, S. H., & Bilsky, W. (1987). Toward a universal psychological structure of human values. *Journal of Personality and Social Psychology, 53,* 550–562.

Schwartz, S. H., & Bilsky, W. (1990). Toward a theory of the universal content and structure of values: Extensions and cross-cultural replications. *Journal of Personality and Social Psychology, 58,* 878–891.

Schwartz, S. H., & Boehnke, K. (2004). Evaluating the structure of human values with confirmatory factor analysis. *Journal of Research in Personality, 38,* 230–255.

Schwartz, S. H., & Huismans, S. (1995). Value priorities and religiosity in four western religions. *Social Psychology Quarterly, 58,* 88–107.

Smith, E. R., & Mackie, D. M. (2007). *Social psychology* (3rd ed.). New York: Psychology Press.

Stern, P. C. (2000). Toward a coherent theory of environmentally significant behavior. *Journal of Social Issues, 56*(3), 407–424.

Stern, P. C., Dietz, T., Abel, T., Guagnano, G. A., & Kalof, L. (1999). A value-belief-norm theory of support for social movements: The case of environmentalism. *Human Ecology Review, 6,* 81–97.

Stevenson, R. (2007). Schooling and environmental/sustainability education: From discourses of policy and practice to discourses of professional learning. *Journal of Environmental Education, 13,* 265–285.

UN. (1987). Report of the United Nations General Assembly. Forty-second session. Item 83 (e) of the provisional agenda. Development and International Economic Co-operation: Environment. Report of the World Commission on Environment and Development. A/42/427. Annex: "Our Common Future".

UNEP. (1972). Report of the United Nations Conference on the Human Environment in Stockholm 1972. http://www.unep.org/Documents.multilingual/Default.asp?DocumentID= 97&ArticleID, Agosto 2006.

UNEP. (2011). *Visions for change: Recommendations for effective policies on sustainable lifestyles.* Nairobi: UNEP. http://www.worldinbalance.net/agreements/1987-brundtland.php, October 2008.

UNEP/LAC-IGWG.XIV/Inf.12 (2003). *Latin American and the caribbean proposed programme on environmental education within the framework of sustainable development.* Fourteenth Meeting of the Forum of Ministers of the Environment of Latin America and the Caribbean Panama City, Panama, November 2003.

UNESCO. (1975). The belgrade charter international workshop on environmental education held in Belgrade, Yugoslavia, October 13–22, 1975. http://portal.unesco.org/education/en/file_ download.php/47f146a292d047189d9b3ea7651a2b98The+Belgrade+Charter.pdf, Agosto 2006.

UNESCO. (1998). World declaration on higher education for the twenty-first century: Vision and action and framework for priority action for change and development in higher education. http://www.unesco.org/education/educprog/wche/declaration_eng.htm, Agosto 2006.

UNESCO. (2004). Report for the higher-level panel meeting on the united nations decade of education for sustainable development (2005–2014): Preparing the draft international implementation scheme, a brief summary of the preparatory process, Paris, July 2004.

UNESCO. (2005). Report by the director-general on the United Nations of education for sustainable development: Draft international implementation scheme and UNESCO's contribution to the implementation of the decade (2005–2014). Hundred and seventy-second session, Paris, August 2005. http://www.unesco.org/education/desd, Agosto 2006.

UNESCO. (2007). *The UN decade of education for sustainable development (desd 2005–2014): The first two years.* Paris: UNESCO.

UNESCO. (2009). *Bonn declaration.* UNESCO World Conference on Education for Sustainable Development, March 31–April 2, 2009. Bonn, UNESCO. http://www.esd-world-conference-2009.org/fileadmin/download/ESD2009_BonnDeclaration080409.pdf.

UNESCO. (2012). *Shaping the education of tomorrow: 2012 full-length report on the UN decade of education for sustainable development.* DESD Monitoring and Evaluation—2012. UNESCO Education Sector.

Von Eckardt, B. (1996). *What is cognitive science?* USA: The MIT Press (1993).

Wals, A. E. J. (2009). A mid-decade review of the decade of education for sustainable development. *Journal of Education for Sustainable Development, 3*(2), 195–204.

Wals, A. E. J. (2014). Sustainability in higher education in the context of the UN DESD: A review of learning and institutionalization processes. *Journal of Cleaner Production, 62*(2014), 8–15.

Worldwatch. (2012). *State of the world 2012: Moving toward sustainable prosperity.* Washington, DC: Worldwatch Institute.

Part I
Socio-psychological Model to Determine Sustainable Behavior

Chapter 2
Theoretical Explanatory Frameworks for Sustainable Behavior

In the previous chapter, the discussion focuses on the definition of sustainable behavior, underlying principles, and the background of education for sustainable development (ESD). This chapter discusses the cognitive theory which models sustainable behavior under the information-processing approach. Additionally, in this section, the most widely known social-psychology models for explaining attitudes which promote the study of sustainable behavior and the factors associated with them are shown. This provides conceptual frameworks which identify the factors explaining sustainable behavior (specifically in situations in which social dilemmas exist, as is the case for many environmental problems and their economic, social, and cultural contexts as indicated in ESD).

2.1 Theoretical Frameworks Which Explain Sustainable Behavior

Virtually all conceptual schemes which have been used in psychology (Chacón-Fuertes 2001) have been applied to explain pro-environmental behavior (PEB) and can be used to elucidate sustainable behavior. Some of the known explanatory frameworks are behaviorism, psychoanalysis, cognitivism, evolutionary psychology, and interdisciplinary systemic approaches, and many variations may be found within each framework.

According to Corral-Verdugo (2001), behaviorists maintain that sustainable behavior, like any behavior, is under control of both external stimuli and individual's circumstances. Behavior is activated shortly after a conditioned stimulus or after a primary reward if no conditioned stimulus exists. The core tools of operant conditioning are positive and negative reinforces. Positive reinforcement is a consequence of a given behavior which causes that behavior to occur with greater frequency. Negative reinforcement or punishment is a consequence of a behavior which causes that behavior to occur with less frequency. A lack of any consequence following a behavior leads to the cessation of that behavior. Whenever a behavior is

© Springer International Publishing Switzerland 2015
M. Juárez-Nájera, *Exploring Sustainable Behavior Structure in Higher Education*,
Management and Industrial Engineering, DOI 10.1007/978-3-319-19393-9_2

inconsequential, producing neither favorable nor unfavorable consequences, it will occur with less frequency. When a previously reinforced behavior is no longer reinforced with either positive or negative reinforcement, it leads to a decline in the response. For behaviorists, no internal phenomenon significantly explains behavior because internal phenomena are intangible and subjective and therefore may not be scientifically studied.

By contrast, cognitive science indicates that internal or mental phenomena lead to behavior. People's knowledge, attitudes, or beliefs are variables which they form based on their interaction with their environment. These may be expressed in the form of ecological habits. Cognitive science is the study of the nature of intelligence and emphasizes algorithms (mathematical operations) intended to simulate human behavior on a computer (Medin et al. 2005; Von Eckardt 1996).

Psychoanalysts see the dichotomy between environmental conservation and environmental degradation as a result of a struggle between creative (Eros) and destructive (Thanatos) impulses of the human unconscious, or between biophilia (love for living systems) and death wishes. Currently, there is a high rate of degradation which would seem to indicate that Thanatos (the destructive) prevails over Eros (the creative), which conforms to Freud's pessimistic explanation of psychological mechanisms of human aggressiveness (Fromm 1973). Although psychoanalysts have offered many proposals to counter the effect of destructive impulses toward the environment, little or no research has been carried out from a psychodynamic perspective to corroborate the relevance of these proposals.

Evolutionary psychology ensures that conservation of the environment and biodiversity can be understood as a necessity for maintaining a safe, high-quality environment and the perpetuation our species. This is useful to understand as we manipulate the environment according survival needs. However, some evolutionary biologists believe that actions toward environmental conservation can be explained by reputation-based models which demonstrate an individual's genetic quality by his ability to look after him/herself (selfishness), his/her family (genetic altruism), or others in hopes of retribution (reciprocal altruism). Helping others at a small cost to oneself is a signal of genetic quality because this characteristic is costly to maintain, and only high-quality individuals can afford the cost. Some evolutionary psychologists argue that altruism evolves into a form of behavior which enables the preservation of the social group and therefore of individuals and their genes. Other evolutionists (Fromm 1973) suggest a human biophilia which is an affinity of our natural love for life and which helps sustain life.

Models which take a systemic approach, by trying to gain further inclusiveness in explaining why people behave in a pro-ecology manner, include effects of situational variables (physical and regulatory contexts) and other variables of an extra-psychological nature (Weisbuch 2000). Some variables included are individual characteristics such as age, sex, social class, income, educational level, or contextual factors such as social norms. Table 2.1 summarizes the explanatory frameworks presented above.

From the frameworks presented in Table 2.1, cognitive science (Medin et al. 2005; Von Eckardt 1996) seems to be the most useful in explaining peoples'

Table 2.1 Explanatory theoretical frameworks for pro-environmental behavior and their fundamental elements (based on Corral-Verdugo 2001)

Theoretical framework	Fundamental elements	Explanation of PEB
Behaviorism Developed by Skinner in 1938	Operant conditioning	PEB is generated and maintained by its positive and immediate consequences
Cognitive psychology "Revolution of cognition" in the sixties	Information processing variant: • Theory of planned behavior • Norm-activation theory • Habit formation • Cognitive dissonance	Individual generates sustainable provisions that are processed, stored, and used in his or her brain and mind
Psychoanalysis Developed by Freud in 1900	Intra-psychic apparatus	In the struggle between Eros and Thanatos (Fromm 1973), there is a predominance of the latter
Evolutionary psychology Based on Darwin's postulates in 1859	Genetic stress variant: • Genes and egoistic individuals • Cooperation and altruism • Altruism and SB • Egoism and SB • Biophilia hypothesis	The effect of PEB is reciprocal altruism that may become disinterested altruism or biophilia (Fromm 1973)
Systemic theories	Interrelated factors	PEB is a product of complex operating effects within systems of relationships between variables

behavior in relation to aspects of their environment, welfare, and material and social safety within society. Cognitive science is an interdisciplinary area with contributors from various fields, including cognitive psychology, which is a branch of psychology according to which investigates internal mental processes such as problem solving, memory, and language. The most relevant school of thought emerging from this approach is known as cognitivism, which characterizes people as dynamic information-processing systems whose internal and mental operations (beliefs, attitudes, or perceptions) might be described in computational terms. The information-processing approach will be presented in the following section.

2.1.1 Information-Processing Approach

The conceptual framework which brings some structure to the pandemonium of contemporary behavior research is cognitive science (Leahey and Harris 2001).

This explanation prevails today (Matthews et al. 2000). The twentieth-century emergence of the conceptual framework of information processing to explain the human cognitive process was mainly due to the rapid development of computer science and the impressive demonstration of artificial intelligence in the late 1950s and formal analysis of cognition in the 1960s. Since then, the dominant theory has been the cognitive information-processing model which Broadbent, among other contributors, put forward. These scientists viewed mental processes as computer software inside hardware, (the brain). They referred to input as information entered into a computer, its representation, computation or processing, and output as new information.

The mind–body problem, and its modern subjective expression called "consciousness," is a topic which has been vehemently debated by philosophers for millennia and more recently by psychologists and biologists (Chacón-Fuertes 2001). The question of whether consciousness plays a role in the production of behavior, or whether it is a powerless observer of the world, and the body's response to behavior, seems to present two competing approaches based on information processing: the symbolic system hypothesis and the connectionist assumption (Leahey and Harris 2001).

The symbolic system hypothesis establishes that the mind is like a computer program. At the core of the program is a manipulation of symbols representing the world through a set of formal rules, analysis of stimuli, and selection of responses. In its simplest form, information arises from the senses and is transformed into an internal representation, and the subject produces an answer (Matthews et al. 2000).

Meanwhile, the connectionist assumption makes no distinction between types of memory. Instead, this approach states that the architecture of cognition consists of multiple simple processing units, very similar to neurons in the interconnected network of the brain. Each unit is identical to all other units, and learning, memory, and thinking are all changing patterns of activity in the network as a whole (Laehey and Harris 2001).

By analogy, the mind represents software or sequences of instructions carried out by computers or other hardware. This software does not refer to a physical machine or hardware. At the most fundamental level, brains resemble computers in their use of binary representations. The fundamental "machine code" of computers is expressed in "zeros" and "ones," and the neurons of the brain are either firing ("on") or resting ("off") (Matthews et al. 2000).

Cognitive information processing between inputs and outputs is more complex. However, the number and nature of intermediate steps depend on the particular approach. That is, the internal structure of processing or the order in which processes operate and how they feed into one another are key elements to understanding existing approaches. Two approaches have been proposed: processing systems which carry out their calculations in series or in parallel. Series models assume that each operation is carried out one step at a time; the last operation must finish before the next one in the series commences, as occurs in a conventional computer program. Parallel models, however, are comprised of multiple processors operating simultaneously. Unlike conventional computers, brains are composed of thousands of massively interconnected simple computation units (neurons) operating simultaneously (Leahey and Harris 2001).

This difference between brains and computers brings up several reasons to doubt the validity of the hypothesis of the serial processing of human cognitive symbolic processing. First, the human brain is capable of thinking and reacting quickly; many computational stages are carried out simultaneously. Secondly, the failure of traditional artificial intelligence to simulate simpler human skills such as recognizing friends' faces, reading, writing, and moving around inside a room full of objects, despite years of work and the increasing possibilities of computers, has led many psychologists to suspect that the serial processing model of the symbolic system in the human mind is incorrect, and instead of looking at the computer as our model for the mind, they should look at the brain (Leahey and Harris 2001).

At present, there is an emerging hypothesis (Leahey and Harris 2001) which could reunite the two approaches of cognition; the human mind is a hybrid of both (Dennett 1995). It is possible that the human mind in its rational aspects is a serial performance processor, especially when thoughts are transformed into awareness. For example, when we think or write, an idea and a thought appear simultaneously. Meanwhile, more automatic and unconscious aspects of the human mind would be of a connectionist nature. Consciousness is a virtual machine installed by socialization in parallel processors in the brain. Socialization nourishes us with language. However, with language, we speak and think one thought at a time, creating a serial processing of consciousness. Humans are flexible creatures who do not change their physical nature, but rather their programs. These programs are cultures that are tailored to places and times. Learning a culture raises awareness and consciousness, and consciousness is an adaptive process because it provides the ability to reflect upon one's own actions, to think about alternatives, plan in advance, acquire general knowledge, and be a member of the society.

The computational framework has attracted a variety of criticisms. First, an assortment of philosophical issues relate to traditional questions such as the mind–body problem. Further controversies concern the experience of consciousness upon the presentation of mental states. Second, the computer metaphor may be broadly correct but unhelpful, because of the diversity of possible computational systems, constructed based on different principles, to explain any given set of data. Conversely, what computers do well—perform high-speed mathematical functions, abide by rule-governed logic—humans do poorly. And what humans do well— form generalizations, make inferences, understand complex patterns, and experience emotions—computers do poorly or not at all. Third, the computer metaphor may be appropriate to some psychological functions, but not to some of the essential attributes of humanity such as emotion, personality, creativity, and intelligence. Leaving these fundamental issues aside, cognitive models may have a surprising range of applications. Nowadays, there is a well-established link between emotional disorders and particular styles of information processing, characterized by negative self-referent cognition and irrational beliefs. Computers do not have feelings, but emotions and personality may nevertheless have a cognitive basis. Furthermore, the computer metaphor suggests undue passivity. Computers run programs fomented by an external agent, while people pursue goals actively and flexibly within complex environments. In other words, the nature of behavior resides in the dynamic interplay between person and environment, rather than in some fixed program.

However, none of these limitations should be considered to be fundamental difficulties for the computer metaphor which has proven to be extremely useful in explaining many areas such as personality, emotional disorders, and human behavior, and has remained the only scientifically acceptable bases for conceptualizing performance (Matthews et al. 2000). Considering the information-processing approach as the conceptual framework to describe human behavior generates the question: Why the need to promote a study of human behavior in a world where the integrity of animal and plant species, as well as the welfare and material security of individuals and society in present and future human generations, is threatened?

The significance lies not only in promoting a study of sustainable behavior, but also in identifying factors which are capable of change. Psychologists and sociologists alike are exploring associated factors in order to understand and produce a model for human behavior which approximates in a transparent manner the current situation across diverse cultural environments.

The following section examines three dominant theoretical frameworks considering those factors which promote or limit individual behavior. These frameworks are instruments which can be helpful in analyzing determinants of sustainable behavior.

2.1.2 Socio-Psychological Attitude-Behavior Models

Contemporary scholars have built complex models of relationships among several key behavioral determinants such as experience, knowledge, beliefs,[1] attitudes,[2] and values.[3] Despite the diversity of specific applications of attitude-related theories,

---

[1] According to Rokeach (1973), a belief is a simple proposition, conscious or not, which may be inferred from what a person says or does, and which may be preceded by the words "I believe that." Any belief consists of three parts: cognitive (knowledge), affective (feeling), and conative (action). The three main categories of belief are as follows: descriptive or existential (I believe that the sun rises in the east), evaluative (I believe that trees are beautiful), and prescriptive or exhortative (I believe that trees must be respected). Beliefs are formed during childhood. The set of beliefs that an individual has regarding the surrounding socio-physical reality is called a belief system.

[2] An attitude is a smaller set of related beliefs. It is also a comprehensive, relatively enduring belief regarding an object or situation which predisposes the person to respond in a certain way to that object or situation (Caduto 1995).

[3] Values are forged from sets of interrelated attitudes. Values are enduring beliefs about a certain behavior or ideal way of life which is personally or socially preferable to an alternative behavior or way of life (Caduto 1995).

they may be separated into two socio-psychological models which take into account factors which promote or limit an individual's behavior (Kaiser et al. 2005).

The two general models are as follows: (a) theory of planned behavior (Ajzen 1991) and (b) norm-activation theory (NAT) (Schwartz 1977). While the first has its basis in deliberation based on rational choice and self-interest, the second is grounded in values and moral norms. Recently formulated, the value-belief-norm framework (VBN) (Stern et al. 1999; Stern 2000) is a generalization of the NAT.

2.1.2.1 Theory of Planned Behavior

The Theory of Planned Behavior (TPB) supposes that behavior is predicted by an individual's intention to perform. In turn, intention is seen as a function of (a) a person's attitude toward this behavior, (b) subjective norms, and (c) people's perceived control, shaped by their estimation of their own strength to perform a behavior which can be prevented (or facilitated) by their abilities or situational factors (Armitage and Conner 2001; Kaiser et al. 2005; Montalvo 2002; Wehn 2003). Figure 2.1 outlines attitudinal relationships of sustainable behavior using the TPB model proposed by Ajzen (1991, 2001, 2005).

There is a great interest in TPB research. Harland (2001), Montalvo (2002), and Wehn (2003) found hundreds of empirical studies based on this model and its predecessor, the theory of reasoned action. Such popularity may be attributed to specificity with which instructions for applying these models were outlined by Ajzen and Fishbein in 1980 and also to the fact that these models are consistent (Harland 2001). TPB has become the most influential attitude-behavior model in socio-psychology and in environmental psychology (Kaiser et al. 2005). In fact, with respect to the environment, health care, nutrition, sports, etc., many studies have found support in (aspects of) TPB.

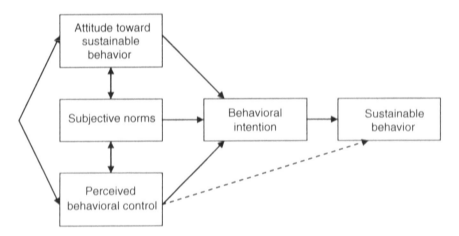

Fig. 2.1 Sustainable attitude model proposed by Ajzen (based on Montalvo 2002)

2.1.2.2 Norm-activation theory

The NAT ascribes a significant role to personal norms. It postulates that personal norms are intrinsically motivated self-expectations with regard to morally appropriate behavior. Personal norms, if activated, are experienced among individuals as feelings of personal obligation, of either denying or not denying the consequences of their behavioral choices regarding the welfare of others. Behavioral expectations stem from personal norms which are grounded within and across individuals, and not from social norms, in a specific social group (Stern et al. 1999; Harland 2001).

The NAT holds that activation of personal norms occurs under the influence of four situational activators and two personality trait activators. The four situational activators are (a) *awareness of need*, or the extent to which a person's attention is focused on the existence of another person or an abstract entity, such as environment, in need, (b) a person's sense of feeling *responsible* for the consequences of the behavior regarding that person's welfare, (c) *efficacy*, which refers to the extent to which persons recognize actions which might alleviate need, and (d) *ability*, or the extent to which one possesses the resources or capabilities needed to perform the action in question. Two personality traits refer to predispositional influences regarding norm activation: *awareness of consequences*, which refers to a person's receptivity for cues signaling situational needs, and *denial of responsibility*, which refers to people's inclination to deny responsibility for the consequences of their behavioral choices directed toward the welfare of others. The four situational activators and the two personality traits determine whether or not a behaviorally specific personal norm becomes activated (Harland 2001; Stern 2000).

The numerous applications of NAT in the environment domain have provided support for several of the relationships proposed in the model (Harland 2001). However, Harland (2001) and Stern (2000) indicate that several authors have noted that in these models, a decisive role has been assigned to personal norms. This view of personal norms raises the question whether the central role assigned to personal norms in NAT is justified in all cases and suggests that personal norms could play a less striking role, as in other models.

2.1.2.3 Value-Belief-Norm Theory

The VBN theory unites the value theory, the norm-activation theory, and the perspective of new ecological paradigm[4] (NEP) through a causal chain of five variables which guide an individual toward behavior: The first latent factor is Schwartz's (1977) set of personal values (altruism, selfishness), traditionalism, and openness to change values; the second factor is the NEP (Dunlap and van Liere 1978; Dunlap et al. 2000); the third and fourth factors take into account the two elements of

---

[4]NEP states that human beings are part of natural world and subject to the same rules which govern nature, such as the interdependence of species (Dunlap and van Liere 1978; Dunlap et al. 2000).

the NAT regarding moral norms, awareness of consequences (AC), and ascription of responsibility (AR) with respect to general conditions of the biophysical environment; and the fifth element includes personal norms for pro-environmental action. This model explains environmental activism, environmental citizenship, support for policies, and behavior in private sphere (Stern et al. 1999; Stern 2000). Previous authors' works support the rationale and empirical causal ordering of factors.

The causal chain starts with central elements, such as relatively stable personality, and belief structures and moves toward beliefs more focused on environment–human relationships, its consequences, and individual responsibility to take corrective actions. Stern (2000) hypothesizes that each variable in the chain directly affects nearby variables and can also directly affect variables which appear later in the chain. Personal norms leading to pro-environmental actions are activated by individuals' belief that environmental conditions threaten things which they value and that they can act to reduce the threat. These norms create a general predisposition which affects many types of behaviors carried out with pro-environmental intention. Additionally, specific personal behavioral norms and social-psychological factors can affect individuals' pro-environmental behavior. Figure 2.2 shows the diagram proposed by Stern et al. (1999).

Stern (2000) recommends that studies which examine only attitudinal factors probably find effects in an inconsistent manner, because effects are contingent on abilities and contexts. Studies which examine only contextual variables such as material incentives, social norms, or the introduction of new technologies may find effects which depend on people's attitudes or beliefs, although the model attributes these effects to other causes. Studies of simple variables demonstrate that a particular theoretical framework has explanatory strength, but they do not contribute much to the comprehensive understanding of individual behaviors which are environmentally significant which are needed to change people's actions.

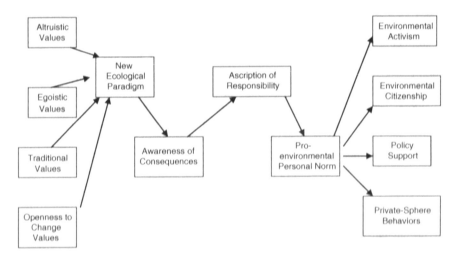

Fig. 2.2 Schematic model of variables in the value-belief-norm theory (based on Stern et al. 1999)

Harland (2001) and Stern (2000) consider that the NAT is an effective tool because they found the attitudinal component to be superior to the normative component in determining the willingness of behavior. This may have been caused by the fact that the normative component of the model is not moderate. On the other hand, Kaiser et al. (2005) compare TPB and VBN: TPB more fully explains proportion of explained variance. More importantly, the adjusted statistics reveal that only TPB appropriately represents the relationships among its concepts, whereas the VBN model does not.

So, which social-psychology model is to be used to determine factors which foster sustainable behavior? Should we accept a model which focuses on rational choice and individual self-interest but which denies moral considerations, or a model based on values and moral norms through its generalization? What philosophical point of view should be considered in morally relevant situations in which social dilemmas are presented—that is, when one's self-interest and the interest of others are contradictory, when there is a tension between individual and collective rationality (social dilemmas; Kollock 1998).

The undertakings of the Decade of Education for Sustainable Development (UNESCO 2004, 2005, 2012) involve social dilemmas: poverty reduction, gender equality, health promotion, environmental protection and conservation, rural transformation, human rights, intercultural understanding and peace, sustainable production and consumption, natural and cultural diversity, and communication and information technology. Several authors (Axelrod 1984, 2004; Felkins 1995; Kollock 1998; Macy et al. 2002; Santos et al. 2006) have analyzed the dynamics of social dilemmas. In general, they point out that agent-based models or models "from the bottom-up" assume the pre-existence of a very different world in which decision-making is equitably distributed on a global scale, where decision-making is locally organized, stemming from multiple local interactions among autonomous interdependent actors. These authors recommend research on the expectations and effects of generalized reciprocity within groups, the transformation of incentive structures, and a greater focus on heterogeneous dynamic models in understanding social dilemmas.

The current study uses a model adapted from the VBN, because the TPB denies moral considerations, and the VBN is a generalization of the NAT. Additionally, Kaiser et al. (2005) and Corral-Verdugo (2001) indicate that on average, 40 % of behavioral variances are predicted by psychological variables. In other words, 60 % of behavioral variance still remains unpredictable. The field of behavioral change requires synthetic theories or models which incorporate other variables and which explain relationships among these new variables, which are used to explain one or more types of behavior.

The following section presents conceptual frameworks considered in this investigation to determine personal and situational variables which influence the behavior of key individuals in higher educational institutions which foster education for sustainability within their professional activities: teaching, research, outreach, and campus management. Secondly, the proposed model which illustrates relations among personality and contextual factors which explain such behavior is presented.

2.2 How to Identify a Model for Sustainable Behavior

Prediction of sustainable behavior is not simple. It appears to involve a number of variables, none of which is likely to operate without interacting with others. Therefore, the development of a model is a difficult task. Several authors in social psychology (Blamey 1998; Corral-Verdugo and Pinheiro 2004; Harland 2001; Hines et al. 1988/1987; Stern 2000) have used (one of several/a set of) viable attitude-behavior models as a means to identify factors which lead to a change in sustainable behavior, or initially pro-environmental behavior.

Some of the models include familiar theories, such as TPB (Kaiser 1998; Wehn 2003; Montalvo 2006, 2002) and NAT (Arbuthnot 1977; Hopper and Nielsen 1991). Some models also consider organizational factors (Shriberg 2002), personal abilities (Allen and Ferrand 1999), context (Corraliza and Berenguer 2000), and habits (Collins 2001), which are other characteristics suitable for explaining behaviors which frequently have significant impacts through non-attitudinal factors. Identification of advantages and disadvantages of behavior seems to be a straightforward way of detecting these determinants (Harland 2001). However, the identification process is complicated because salient advantages and disadvantages of behavior seem to depend on the perspective from which they are evaluated.

For example, what brings a teacher to introduce in his/her course the values of sustainable development? What motivates students to dispose waste in proper containers? What guides a researcher to develop a project to solve local social problems? What makes staff buy more environmentally friendly goods in order to reduce environmental impact? What guides authorities of higher educational institutions to implement policies to improve the sustainability of their operations or educational context?

The above questions, it would seem then that efforts to explain advantages and disadvantages of behavior need to focus on various factors, such as beliefs, attitudes, motives, and abilities of individuals' to perform, social pressure exerted, moral values at election of acting, individuals' decisions on short- or long-term, socio-demographic conditions, and contextual influences which foster or impede a particular behavior. As well as areas where we want to influence people's behavior and the conceptual framework where contemporary behavioral investigation is based on.

2.2.1 Factors Explaining Sustainable Behavior

The appropriate question concerning sustainable behavior is: What factors are important to foster it and why? In order to prepare the proposed model, a number of conceptual frameworks were researched which provide important considerations in identifying psychological, situational, and contextual factors explaining behavior.

The first theoretical framework is the meta-analysis from Hines et al. (1987) which addresses responsible environmental behavior. This study remains a benchmark for conclusions on behavioral variables.

The second model, the value-belief-norm (Stern et al. 1999) framework, states that, according to values, behavior may be predicted. This model offers an array of five causal factors which determine actions toward social movements.

Thirdly, the theory of multiple intelligences (TMI), developed by Howard Gardner in 1983 and updated in 1993, establishes seven skills (linguistic, logical-mathematical, musical, spatial, body-kinesthetic, and interpersonal and intrapersonal intelligence) which human beings perform in any culture in which they live and grow up. TMI is developed under a distributed vision, that is, inherent to individuals and artifacts that surround them.

The fourth and final theory consists of five psychological dimensions proposed by Corral-Verdugo and Pinheiro (2004) to achieve sustainable actions: effectiveness, deliberation, anticipation, solidarity, and austerity.

The author of this study considers that the norm-activation framework may be a structural descriptive model that aims to gain understanding of the predispositional factors by looking at the structural relationship of the possible determinants of behavior. So, the elements drawn from the conceptual frameworks presented, the psychological and situational variables, causal arrangement of factors which determine an action in favor of the common good, personal skills applied in any culture, and the ideas behind sustainable actions are all part of the notion of sustainability in human behavior.

2.2.1.1 Hines, Hungerford, and Tomera's Model

The model proposed by Hines et al. (1988/1987) identifies four factors which explain elements of willingness to perform an individual process: (1) recognition of the problem as a prerequisite for action, (2) knowledge of the courses of action which are available and most effective in a given situation, (3) the ability to implement strategies of action items, and (4) appropriate knowledge. These factors allow individuals to take action.

Abilities alone are not sufficient to lead to action. In addition, an individual must possess a desire to act. One's desire to act appears to be affected by a host of personality factors. These include locus of control,[5] attitudes, and personal responsibility. Thus, an individual with an internal locus of control, with positive attitudes toward the environment and toward taking action, and with a sense of obligation toward the environment will likely develop a desire to take action.

---

[5]The locus of control represents an individual's perception of whether he/she has the skills to provoke changes through his/her own behavior. External locus of control refers to concepts based on the belief of some individuals do not intend to provoke change, because they attribute change to randomness or other powerful forces (God, government, and father). In the internal locus of control, on the other hand, individuals believe that their activities will likely have an impact.

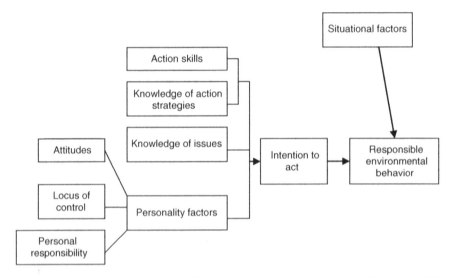

Fig. 2.3 The proposed model of responsible environmental behavior (based on Hines et al. 1988/ 1987)

One remaining category exists which can interrupt this pathway to action: (5) situational factors. Situational factors such as economic constraints, social pressures, and opportunities to choose different actions may enter into the picture and serve either to counteract or to strengthen the variables in the model. For example, if an individual has the cognitive ability, desire, and opportunity to help stop pollution by contributing to a local toxic waste fund, but simply cannot afford to do so, that person will not engage in the environmental action, and in this instance, the model's main pathway will not be followed. Situational factors include age, income, education, and gender. Figure 2.3 presents the model's factors.

This model indicates several areas which are amenable to change by the efforts of environmental educators. The knowledge and skill components, and perhaps the personality components of the model, may be affected through the efforts of educators. Approaches which address both affective and cognitive experiences and which provide individuals with opportunities to develop and practice those skills necessary for environmental action must be developed and implemented in educational systems.

2.2.1.2 Theoretical Framework by Stern et al.

The theoretical framework proposed by Stern et al. (1999), the so-called value-belief-norm theory, explains political activism which is essential to the success of social movements, which seek collective well-being. In some cases, the benefit is distributed among a small and easily identifiable group, but in others, collective benefits are often provided on a local, national, and global scale. This suggests that although some individuals may expect enough personal gain to justify working toward the collective good on egotistical grounds, most are also motivated by a broader, altruistic concern, a willingness to take action even in the face of the "Free Rider Problem" as explained in the "The Tragedy of the Commons"[6] (Hardin 1968), "Voter's Paradox"[7] (Felkins 1995), or "Prisoner's Dilemma"[8] (Poundstone 1992; Axelrod 1984, 2004).

Stern et al. (1999) find that in the USA, many social movements, including the environmental movement, advocate the public good with reference to altruistic values. Such movements work to activate personal norms tied to those values. It is also possible, however, for a social movement to try to activate personal norms based on other types of values. For example, some conservative social movements, which see traditional values of duty, family loyalty, and the like as essential for providing public benefit such as social order, refer to these values in attempting to activate feelings of personal obligation to support the movement's objectives.

Stern et al. (1999) propose that norm-based action flow from three factors: (a) acceptance of particular personal values, the personal belief that everything important according to those values is under threat, (b) the belief that actions initiated by the individual can help alleviate the threat, and (c) the belief that these actions will restore the values under threat.

Each of these three factors involves a generalization of Schwartz's theory (1977): (1) The original theory presumes altruistic values exist. The revised, broadened theory holds that personal norms may have roots in other values as well as in altruistic values and those levels of altruism and other relevant values may vary across individuals. (2) The original theory emphasizes awareness of adverse consequences of events for other people; the broadened theory emphasizes threats to whatever objects are the focus of the values that underlie the norm. (3) Norm activation depends on ascription of responsibility to oneself for the undesirable consequences to others; the broadened theory emphasizes beliefs regarding responsibility for causing undesirable effects or the ability to alleviate threats to any valued object.

In expanding the range of valued objects to be given theoretical consideration, Stern et al. (1999) adopt the typology of value developed by S.H. Schwartz (Schwartz and Blisky 1987, 1990; Schwartz 1994; Schwartz and Huismans 1995;

---

[6]The Tragedy of the Commons describes conflicts between individual and group interest through an example of a common pasture shared by the local community with free access and no restrictions. Every individual realizes that his interest is best served by bringing as many cattle as possible to the pasture although the fodder is limited and it is obvious that if everyone does so, the common goods will be completely exhausted.

[7]The Voter's Paradox describes conflicts between individual and group interest in situations where, for instance, a person votes or volunteers in situations where collective action is involved, and people really cooperate, but they do (so) by self-interest.

[8]The Prisoner's Dilemma describes a model of cooperation between two or more individuals (or corporations, or countries) in ordinary life in which, in many cases, it would be personally worthwhile for each individual to not cooperate with the others (better to desert).

Schwartz and Boehnke 2004). It is worthy to stress some general considerations under value conceptual framework.

Universal Aspects of Human Values

Values are forged from sets of interrelated attitudes. Values are enduring beliefs about a certain behavior or ideal way of life which is personally or socially preferable to an alternative behavior or way of life (Caduto 1995). According to Caduto (1995), values associated with a particular behavior are called instrumental values (e.g., honesty, respect for the environment) and those involving ideal ways of life are called final values (e.g., peace in the world, environmental quality).

According to Pereira de Gómez (1997), values are classified into *physical* (e.g., health, physical ability, and self-awareness), *intellectual* (e.g., attitude toward scientific knowledge, thought, and critical consciousness (criticism)), *aesthetic* (e.g., sense of beauty, respect for different artistic expressions), *ethical* (e.g., honesty, kindness, truth, justice, tolerance), *socio-emotional* (e.g., sense of belonging, awareness of others, solidarity, democracy, brotherhood, service), *religious* (e.g., knowledge of one's mission and living accordingly, recognition of one's limitations or deference to a higher power), and *liberty* (e.g., convictions, capacity to analyze, openness to pluralism, human rights).

According to Schwartz (1994), values have five conceptual aspects: A value is a belief pertaining to desirable end state or modes of conduct, that transcends specific situations, guides selection or evaluation of behavior, people, and events, and is ordered according to importance relative to other values to form a system of value priorities.

Implicit in this definition of values as goals is that (1) they serve the interest of some social entity, (2) they can motivate action (giving it direction and emotional intensity), (3) they function as standards for judging and justifying action, and (4) they are acquired both through socialization to dominant group values and through the unique learning experiences of individuals.

In order to cope with reality in a social context, groups and individuals cognitively transform the necessities inherent in human existence and express them in the language of specific values about which they can then communicate. Specifically, values represent, in the form of conscious goals, responses to three universal requirements with which all individuals and societies must cope: (1) needs of individuals as biological organisms, (2) requisites of coordinated social interaction, and (3) requirements for the smooth functioning and survival of groups. Ten motivationally distinct types of values were derived, evaluated, and confirmed to be recognized within and across cultures: power, achievement, hedonism, stimulation, self-direction, universalism, benevolence, tradition, conformity, and security. The ten value types (see Appendix A) are grouped in a semicircular structure under four categories: self-enhancement, openness to change, self-transcendence, and conservation. Figure 2.4 depicts the complete pattern of relations among values postulated by the theory.

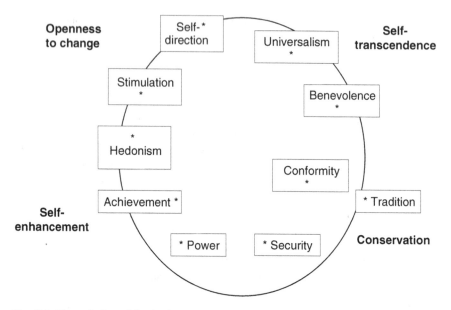

Fig. 2.4 Theoretical model of relations among 10 motivational types of values (based on Schwartz and Boehnke 2004)

The most important feature of value theory is the structure of dynamic relationships among 10 values. According to the theory, expressive actions of any value have practical, psychological, and social consequences which may create conflict or be compatible with the search for other values. For example, actions which express values of hedonism are likely to be in conflict with those which express values of tradition, or acting on values of self-direction is likely to conflict with values of conformity. On the other hand, values of hedonism are compatible with values of self-direction; values of tradition are compatible with values of conformity. Schwartz's (1994) study in 44 countries and his study conducted in 2004 in 27 countries reveal systemic associations of many behaviors, attitudes, and personality variables with priorities for these values. The circular arrangement of values represents a continuous motivational. The closer the two values are in any direction around the circle, the greater the similarity of their underlying motivations.

The ten types of values are listed in the first column of Table 2.2, each defined in terms of the central goal of that category of values. The second column lists 45 specific values as primary examples representing each type.

The theory sustains that there are 10 core values identifiable in all societies, and these values can be arranged to form a semicircular structure based on inherent conflicts or compatibility between the motivational goals implicit to these values.

The conceptual framework proposed by Stern et al. (1999) states that behavior may be predicted according to one's values. This model offers an array of five causal factors which determine actions toward social movements. Also, it extends

Table 2.2 Motivational types of values (based on Schwartz 1994; Schwartz and Boehnke 2004)

Definition	Example of values
Power	Social power, control over others, dominance Health Authority, the right to lead or command Preserving public image
Achievement	Ambitious, wealth, material possessions, money Influential, having an impact on people or events Capable Successful
Hedonism	Pleasure Enjoying life Self-indulgent
Stimulation	An exciting life, stimulating experiences A varied life, filled with challenge, novelty, and change Daring
Self-direction	Freedom Creativity Independent Choosing own goals Curious, interested in everything, exploring
Universalism	Equality, equal opportunities for all A world of peace, free of war and conflict Unity with nature, fitting into nature Social justice, correcting injustice, care of the weak Broad-minded Preventing and protecting pollution, conserving natural resources A world of beauty
Benevolence	Responsible Loyal, true friendship, faithful to friends Honest, genuine, sincere Amiable Forgiving, willing to pardon others
Tradition	Respecting the earth, harmony with other species Moderate Humble Accepting portion in life Devote
Conformity	Politeness Self-discipline, self-restrain, resistance to temptations Honoring parents and elders, showing respect Obedient, dutiful, meeting obligations
Security	Social order National security Reciprocation of favors Family security, safety for loved ones Clean

considerations of the activation of moral norms not only to environmental issues, but also to economic, social, and cultural issues implicit in the concept of sustainability.

2.2.1.3 Howard Gardner's Theoretical Framework

The TMI points out the theoretical framework in relation to the range of skills deployed by human beings across all cultures. Gardner (2001) states that human cognition according to Piaget's concepts (Pansza 1999; Salles 1999) or actual cognitive science must include a repertoire of skills more universal and more comprehensive than they are now.

In order to formulate the TMI, Gardner (2001) reviewed the literature using eight criteria or "signs" to define intelligence: (1) potential isolation from due to brain damage, (2) the existence of idiot savants, prodigies, and other exceptional individuals, (3) an identifiable core operation or set of mental operations, (4) an individual's distinctive development history, along with a definable set of "end-state" performances, (5) an evolutionary history and evolutionary plausibility, (6) support from experimental psychological tasks, (7) support from psychometric findings, and (8) the individual's ability to decode a symbolic system. Howard Gardner views intelligence as the capacity to solve problems or fashion products which are valued in one or more cultural settings. This definition tells us nothing about the sources of such capabilities or the means of measuring them. Perhaps many of these skills do not lend themselves to measurement by verbal methods which largely depend on a combination of logic and language skills.

Based on this definition, and relying on a range of the above criteria and prerequisites, Gardner initially formulated a list of seven types of intelligence: (1) linguistic, (2) logical-mathematical, (3) musical, (4) spatial, (5) body-kinesthetic, (6) personal intelligence directed toward others (inter), and (7) personal intelligence directed toward oneself (intra). The TMI (Gardner 2001) establishes seven skills which human beings perform in any culture in which they live and grow up. TMI is developed under a distributed vision, that is, inherent to individuals and artifacts that surround them. In other words, intelligence does not end with the skin, but rather encompasses tools (paper, pencil, and computer), documentary memory (contained in files, notebooks, and diaries), and a network of acquaintances (coworkers, colleagues, and other persons to whom one communicates by telephone or electronically). In addition, Gardner considers how skills may be put to use in a diverse range of educational settings (Gaxiola 2005).

Gardner claims that the seven types of intelligence rarely operate independently. They are used simultaneously and tend to complement each other as people develop skills or solve problems. Human beings are organisms who possess a basic, uniquely blended set of intelligences. These intelligences are amoral—they may be put to constructive or destructive use. However, leaders, or people with skills which cross boundaries among intelligences, can affect other people emotionally, socially, and cognitively. They link individuals from different intellectual trends, scopes (disciplines, professions), and fields (people, institutions, award mechanisms, and everything which makes it possible to judge the quality of staff performance in a large enterprise).

Table 2.3 shows the relationships among seven types of intelligence identified by Gardner: linguistic, logical-mathematical, musical, spatial, body-kinesthetic, inter-personal, and intrapersonal. The table also presents their channel of access in humans and their neural representation (from a descriptive process) and examples of the most representative profile of what type of people exhibit for each type of intelligence.

Applying this theory to educational contexts, several criticisms arise with respect to Howard Gardner's conceptualization of multiple intelligences. However, this theory holds that (1) multiple intelligences act on a value system whereby students with a diversity of abilities can learn and succeed; (2) that learning is exciting and that hard work by teachers is necessary; (3) that the exchange of constructive suggestions and formal and informal ideas embedded in the curriculum and the evaluation of educational activities are valid for the students, as well as for the broader culture; (4) that the arts may be employed in order to develop people's abilities and comprehension within and across disciplines; and (5) that multiple intelligences are means to fostering high-quality student work. These features are highly pursued in education for sustainability.

2.2.1.4 Psychological Dimensions by Corral-Verdugo and Pinheiro

With respect to psychological factors which affect or are affected by the interaction between the individual and the environment and the lack of clarity in dimensions behind the definition proposed for sustainable behavior (see Chap. 1), and with the goal of complying with that idea, given that individual and group behaviors involve social, political, economic, and environmental impacts, the author of this study uses psychological dimensions reported by Corral-Verdugo and Pinheiro (2004).

According to Corral-Verdugo (2010) and Corral-Verdugo and Pinheiro (2004), sustainable behavior should meet at least five psychological features: (1) effectiveness, (2) deliberation, (3) anticipation, (4) solidarity, and (5) austerity. Effectiveness implies swift reaction to requests or demands of the physical or social environment, while deliberation means that behavior must occur with the specific intent of caring and promoting the welfare of humans and other organisms in the environment. Anticipation means that even if one performs a behavior in the current moment, the individual temporarily separates him/herself and projects the action to the future, which is the time to which the current behavior is directed. Solidarity is expressed as the sum of altruistic tendencies and actions deployed in response to concern for others. Finally, austerity raises the need to lead a lifestyle in which consumption of goods and natural resources is limited to that which is necessary, avoiding wastefulness.

Table 2.3 Relationships among types of intelligences and their neuronal representation (based on Gardner 2001)

Kind of intelligence	Channel of access	Neuronal representation	Performance profile
Linguistic intelligence (involves sensitivity to spoken and written language, the ability to learn languages, and the capacity to use language to accomplish certain goals. This intelligence includes the ability to effectively use language to express oneself rhetorically or poetically, and language as a means to recall information)	Oral-auditory tract	Left temporal lobe	Poets, writers, politicians, lawyers, speakers
Musical intelligence (involves skill in the composition, performance, and appreciation of musical patterns. It encompasses the capacity to recognize and compose musical pitches, tones, and rhythms)	Oral-auditory tract	Right hemisphere. Back portions of right brain	Musicians, composers
Logical-mathematical intelligence (consists of the capacity to analyze problems logically, carry out mathematical operations, and investigate issues scientifically, and entails the ability to detect patterns, reason deductively, and think logically)	Visual	Both hemispheres: Left hemisphere has the ability to read and produce mathematical signs, while right hemisphere seems to understand relationships and numerical concepts	Scientists, mathematicians
Spatial intelligence (involves the potential to recognize and maneuver in open spaces and confined areas)	Spatial visual	Back portions of right hemisphere	Sculptors, mathematicians, topologists

(continued)

Table 2.3 (continued)

Kind of intelligence	Channel of access	Neuronal representation	Performance profile
Body-kinesthetic intelligence (entails the potential of using one's whole body or parts of the body to solve problems. It is the ability to use mental aptitudes to coordinate bodily movements)	Visual	Cerebral cortex, thalamus, basal ganglia	Dancers, swimmers, gymnasts
Inter-personal intelligence (concerned with the capacity to understand the intentions, motivations, and desires of others. It allows people to work effectively with others)	Symbolization provided by culture as rituals and religious and mythical systems	Frontal cortex. Front lobes were networks of nerve representing internal environment of individuals converge (feelings, motivations, and subjective knowledge) with the system representing external environment: vision, sounds, tastes, and customs transmitted through the senses	Educators, salespeople, counselors, religious leaders, artists
Intra-personal intelligence (entails the capacity to understand oneself, to appreciate one's feelings, fears, and motivations)			Magicians, warriors, shamans, fortune-tellers

The requirements for sustainability include challenges imposed by the environment (lack of resources, climatic adversity, environmental and social opportunities), and regulatory requirements of social groups (conventions, rules and laws for environmental protection, rules of solidarity, public policies). In addition, individual dispositions (attitudes, beliefs, perceptions, and values) generate conditions in individuals which lead them to act responsibly toward themselves, the environment, and fellow humans.

References

Ajzen, I. (1991). The theory of planned behavior. *Organizational Behavior and Human Decision Process, 50*, 179–211.

Ajzen, I. (2001). Nature and operation of attitudes. *Annual Review of Psychology, 52*, 27–58.

Ajzen, I. (2005). *Attitudes, personality and behavior* (2nd ed.). London: McGraw-Hills.

Allen, J. B., & Ferrand, J. L. (1999). Environmental locus of control sympathy, and pro-environmental behavior: A test of Geller's actively caring hypothesis. *Environment and Behavior, 31*, 338–353.

Arbuthnot, J. (1977). The role of attitudinal and personality variables in the prediction of environmental behavior and knowledge. *Environment and Behavior, 9*, 217–232.

Armitage, C. J., & Conner, M. (2001). Efficacy of the theory of planned behavior: A meta-analytic review. *British Journal of Social Psychology, 40*, 471–499.

Axelrod, R. (1984). *The evolution of cooperation*. New York: Basic Books.

Axelrod, R. (2004). *La Complejidad de la Cooperación. Modelos de cooperación y colaboración basados en los agentes*. México: Fondo de Cultura Económica.

Blamey, R. (1998). The activation of environmental norms, extending Schwartz's model. *Environment and Behavior, 30*, 676–708.

Caduto, M. J. (1995). *Guía para la enseñanza de valores ambientales*. 3 Edición, Madrid: UNESCO.

Chacón-Fuertes, P. (Ed.). (2001). *Filosofía de la Psicología*. Madrid: Biblioteca Nueva.

Collins, C. (2001). Psychological and situational influences on commuter transport mode choice. Report. Melbourne, Australia.

Corral-Verdugo, V. (2010). *Psicología de la sustentabilidad: un análisis que nos hace pro ecológicos y pro sociales*. México: Trillas.

Corral-Verdugo, V., & Pinheiro, J. (2004). Aproximaciones al estudio de la conducta sustentable. *Medio Ambiente y Comportamiento Humano, 5*(1 y 2), 1–26.

Corral-Verdugo, V. (2001). *Comportamiento Proambiental. Una introducción al estudio de las conductas protectoras del ambiente*. Canarias: Editorial Resma.

Corraliza, J. A., & Berenguer, J. (2000). Environmental values, beliefs, and actions. A situational approach. *Environment and Behavior, 32*, 832–848.

Dennett, D. (1995). *La conciencia explicada: Una teoría interdisciplinar*. Barcelona: Paidós.

Dunlap, R. E., & van Liere, K. D. (1978). The new environmental paradigm. *Journal of Environmental Education, 9*, 10–19.

Dunlap, R. E., Van Liere, K. D, Mertig, A. G., & Jones, R. E. (2000). Measuring endorsement of the new environmental paradigm. A revised NEP scale. *Journal of Social Issues, 56*(3), 425–442.

Felkins, L. (1995). The Voter's Paradox. An introduction to the theory of social dilemmas. *The Ethical Spectacle*, Vol. I, No. 9 September.

Fromm, E. (1973). *The Anatomy of Human Destructiveness*. New York: Holt Rineheart.

Gardner, H. (2001). *Estructuras de la Mente. La teoría de las inteligencias múltiples*. 2ª. Edición. México: Fondo de Cultura Económica.

Gaxiola, P. (2005). *La inteligencia emocional en el aula*. México: SM de Ediciones.

Hardin, G. (1968). The tragedy of the commons. *Science*, Vol. 162; 13 December, 1968.

Harland, P. (2001). *Pro-environmental Behavior*. Doctor Thesis. Faculteit der Wiskunde en Natuurwetenschappen en die der Geneeskunde, Universiteit Leiden, the Netherlands.

Hines, J. M., Hungerford, H. R., & Tomera, A. N. (1987). Analysis and synthesis of research on responsible environmental behavior: A meta-analysis. *Journal of Environmental Education, 18* (2), 1–8.

Hooper, J. R., & Nielsen, J. M. (1991). Recycling as altruistic behavior. Normative and behavioral strategies to expand participation in a community recycling program. *Environment and Behavior, 23*(2), 195–220.

Kaiser, F. G. (1998). A general measure of ecological behavior. *Journal of Applied Social Psychology, 28*(5), 395–422.

Kaiser, F. G., Hübner, G., & Bogner, F. (2005). Contrasting the theory of planned behavior with the value-belief-norm model in explaining conservation behavior. *Journal of Applied Social Psychology, 35*(10), 2150–2170.

Kollock, P. (1998). Social Dilemmas: The anatomy of cooperation. *Annual Review Sociology, 1998*(24), 183–214.

Leahey, T. H., & Harris, R. J. (2001). *Aprendizaje y cognición*. 4a. Edición. Madrid: Prentice Hall.

Macy, M. M., & Flache, A. (2002). Learning dynamics in social dilemmas. *PNAS, 99*(3), 7229–7236.

Matthews, G., Davis, R., Westman, S. J., & Stammers, R. (2000). *Human performance: Cognition, stress and individual differences*. Sussex, Great Britain: Psychology Press.

Medin, D. L., Ross, B. H., & Markman, A. B. (2005). *Cognitive psychology* (4th ed.). New York: Wiley.

Montalvo, C. (2002). *Environmental policy and technological innovation, why do firms adopt or reject new technologies?* Cheltenham: Edward Elgar.

Montalvo, C. (2006). What triggers change and innovation? *Technovation, 26*(2006), 312–323.

Pansza, M. (1999). Una aproximación a la epistemología genética de Jean Piaget. *Revista Psicología.* Nov.–Dic., pp. 23–32.

Pereira de Gómez, M. N. (1997). *Educación en valores. Metodología e innovación educativa.* México: Trillas.

Poundstone, W. (1992). *Prisoner's Dilemma* (pp. 215–217). New York: Doubleday.

Rokeach, M. (1973). *The nature of human value.* New York City: Free Press.

Salles, M. (1999). El desarrollo cognitivo—Las aportaciones de Piaget y la Escuela de Ginebra. *Revista Psicología.* Nov.–Dic., pp. 3–8.

Santos, F. C., Pacheco, J. M. & Lenaerts, T. (2006). Evolutionary dynamics of social dilemmas in structure heterogeneous population. *PNAS, 103* (9), 3490–3494.

Shriberg, M.P. (2002). *Sustainability in U.S. Higher Education: Organizational factors influencing campus environmental performance and leadership.* A dissertation for the degree of Doctor of Philosophy (Natural Resources and Environment). The University of Michigan, USA.

Schwartz, S. H., & Boehnke, K. (2004). Evaluating the structure of human values with confirmatory factor analysis. *Journal of Research in Personality, 38*, 230–255.

Schwartz, S. H., & Huismans, S. (1995). Value priorities and religiosity in Four Western religions. *Social Psychology Quarterly, 58*, 88–107.

Schwartz, S. H. (1994). Are there universal aspects in the structure and contents of human values? *Journal of Social Issues, 50*, 19–45.

Schwartz, S. H., & Bilsky, W. (1990). Toward a theory of the universal content and structure of values: Extensions and cross-cultural replications. *Journal of Personality and Social Psychology, 58*, 878–891.

Schwartz, S. H., & Bilsky, W. (1987). Toward a universal psychological structure of human values. *Journal of Personality and Social Psychology, 53*, 550–562.

Schwartz, S. H. (1977). Normative Influences on Altruism. In L. Berkowitz (Ed.), *Advances in experimental social psychology* (Vol. 10, pp. 221–279). New York City: Academic Press.

Stern, P. C. (2000). Toward a coherent theory of environmentally significant behavior. *Journal of Social Issues, 56*(3), 407–424.

Stern, P. C., Dietz, T., Abel, T., Guagnano, G. A., & Kalof, L. (1999). A value-belief-norm theory of support for social movements: The case of environmentalism. *Human Ecology Review, 6*, 81–97.

UNESCO. (2004). Report for the higher-level panel meeting on the united nations decade of education for sustainable development (2005–2014): Preparing the draft international implementation scheme, a brief summary of the preparatory process, Paris, July 2004.

UNESCO. (2005). Report by the director-general on the United Nations of education for sustainable development: Draft international implementation scheme and UNESCO'S contribution to the implementation of the decade (2005–2014). Hundred and seventy-second session. Paris, August 2005. http://www.unesco.org/education/desd, Agosto 2006.

UNESCO. (2012). Shaping the education of tomorrow: 2012 full-length report on the UN decade of education for sustainable development. DESD Monitoring and Evaluation—2012. UNESCO Education Sector.

Von Eckardt, B. (1996) *What is Cognitive Science?* USA: The MIT Press. (1993).

Wehn, U. (2003). *Mapping the determinants of spatial data sharing.* England: Ashgate.

Weisbuch, G. (2000). Environment and institutions: A complex dynamical systems approach. Special Issue: The Human Actor in Ecological-Economic Models. *Ecological Economics, 34*, 381–391.

Chapter 3
Research Method

This chapter presents the research method used to answer the research questions, modification made to norm-activation theory based on value-belief-norm theory, the connection of the theoretical approach with the research design, a justification for the selected case study, a model developed to describe sustainable behavior, and the criteria for data analysis. It also includes a description of how the research was conducted, how the data were collected, and the criteria for the sample selection.

3.1 The Research Strategy

The research questions that originated this work were as follows: (1) Is it possible to reconcile the individual interest with the social interest? (2) If so, under what conditions? and (3) If not, why not? These questions contained a number of hidden questions both at a theoretical and methodological level. Some of the main questions of this work were as follows: What are the factors that determine the development of sustainable behavior? How can we account for and organize those determinants? How can we rate and rank them? How can the degree of conflict between the interest of individuals and the higher educational institution be measured? How can the research strategy approach be validated?

To answer these research questions, the research strategy required a methodology strongly grounded in theory. In a broad sense, the research strategy adopted can be dived into two stages. The first stage refers to the proposition of the norm-activation theory as theoretical framework that integrates several bodies of theory. Chapter 2 mentioned conceptual schemes used to model behavior, with emphasis on cognitive theory on the information processing approach. It also presented the most influential socio-psychological frameworks which take into account limiting or promoting factors of human behavior. The second stage is composed of the fieldwork period used to collect the data and the statistical validation process of the proposed model to assess the willingness of HEI participants to develop sustainable behavior.

© Springer International Publishing Switzerland 2015
M. Juárez-Nájera, *Exploring Sustainable Behavior Structure in Higher Education,*
Management and Industrial Engineering, DOI 10.1007/978-3-319-19393-9_3

3.2 Modifications to the Value-Belief-Norm Model

The proposed model depicted in Fig. 3.1 is based on the value-belief-norm theory (Stern et al. 1999) which is a structural descriptive model that aims to gain understanding of the predispositional factors by looking at the structural relationship of the possible determinants of behaviors.

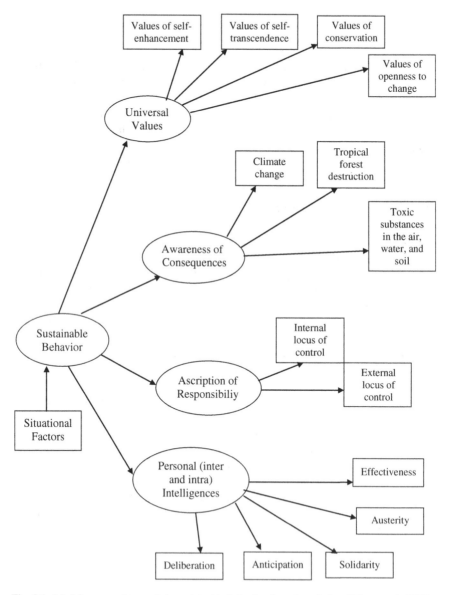

Fig. 3.1 Model proposed to explain sustainable behavior (based on Juárez-Nájera et al. 2010)

As mentioned in the previous chapter, the proposed SB model took into account Schwartz's values theory (1977, 1994) which includes four broad categories of values: self-enhancement, self-transcendence, conservation, and openness to change, as well as 10 types of values distributed along a semicircular structure as shown in Fig. 2.4. Twenty-one specific values were arranged according to topics covered by the Decade of Education for Sustainable Development (see Appendix A for the complete list). Table 3.1 shows each value type associated with 7 DESD

Table 3.1 Twenty-one values selected from the author's larger list (based on Schwartz 1994; Schwartz and Boehnke 2004) and the themes of the decade of education for sustainability

Kind of values	Values of example	Themes of the decade of ESD
Power	Social power, control over others, dominance[1] Health Authority, the right to lead or command[1]	Human rights Health promotion Human rights
Achievement	Ambitious, wealth, material possessions, money[1] Influential, having an impact on people and events[1]	Sustainable production and consumption Human rights
Hedonism	Enjoying life[1]	Human rights and sustainable production and consumption
Stimulation	Varied life, filled with challenge, novelty, and change[1]	Sustainable production and consumption
Self-direction	Creativity Choosing own goals	Technology of information and communication Sustainable production and consumption
Universalism	Equality, equal opportunities for all A world of peace, free of war and conflict Unity with nature, fitting into nature Social justice, correcting injustice, care for the weak Broadminded Prevention and protection of the environment, conservation of natural resources	Gender equality Human rights Rural transformation Poverty alleviation Technology of information and communication Conservation and protection of environment
Benevolence	Responsible	Human rights
Tradition	Respecting the earth, harmony with other species Moderate Accepting one's portion in life	Rural transformation Sustainable production and consumption Intercultural understanding and peace
Conformity	Self-discipline, self-restraint, resistance to temptations	Sustainable production and consumption
Security	Social order	Human rights

[1]Indicates a question regarding an attitude which was inverted upon creating the scales. That is, it is contrary to the underlying principles of ESD

themes: human rights, health promotion, sustainable production and consumption, gender equality, information and communication technology, rural transformation, and inter-cultural understanding and peace; while Stern et al.'s (1999) model arranged 23 values into four broader value categories: altruism, traditional, self-interest, and openness to change values.

The proposed SB model changed two elements in the Stern et al.'s (1999) model. Neither Drake's cultural items nor the new ecological paradigm issue was included. Kaiser et al. (2005), upon comparing VBN and TPB, found that VBN is imprecise with respect to the substantial residual values of NEP (Dunlap and van Liere 1978; Dunlap et al. 2000; Hawcroft and Milfont 2010). They view NEP as inadequately integrated into the model. Furthermore, Stern et al. (1999) found low reproducibility in Drake's cultural items.

Stern et al.'s (1999) proposed model considers two personality traits: awareness of consequences and ascription of responsibility; both are part of Schwartz's norm-activation theory (1977). Hines et al. (1987) considered ascription of responsibility or locus of control to be important factor in their model for determining behavior. These authors considered that the personality variables explain part of the responsible environmental behavior.

Table 3.2 presents both personality traits as they appear in the final question-naire. In the first section, awareness of consequences appeared with respect to three main problems: climate change, tropical forest destruction, and toxic substances in the air, water, and soil. The questionnaire asked whether each problem was very serious, somewhat serious, or will not really be a problem for one-self and one's family, for the whole country, and for other plant animal species. In the second section, ascription of responsibility appeared in three questions related to an internal locus of control and six to an external control. The internal locus of control includes questions concerning oneself and the external locus of control (includes) questions concerning the government and businesses as external supreme entities.

Table 3.3 grouped the intrapersonal intelligences under self-knowledge and self-management categories and the interpersonal intelligences under understanding of others and social skills categories from the emotional competencies by Boyatzis et al. (2002). Twenty out of 72 items were selected among the 10 competency types: self-confidence, emotional self-control, integrity, adaptability, achievement orientation, initiative, empathy, leadership, catalyst for change, and catalyst for teamwork (see Appendix B for the complete list). The final 20 items were analyzed under the underlying principles of the so-called five psychological dimensions which are the actions toward sustainable behavior, namely effectiveness, austerity, solidarity, anticipation, and deliberation; from each competence type, two emo-tional competences for each psychological dimension were chosen.

Finally, eight demographical items were considered in the proposed SB model; some of which Hines et al. (1987) mentioned in their studies. These included income, gender, age, predominant activity related to education (student, faculty, or administrator), and educational level of respondent; also religious denomination as suggested by Schwartz and Huismans (1995).

Table 3.2 Awareness of consequences and ascription of responsibilities (based on Stern et al. 1999)

Awareness of consequences

(1a) In general, do you think that climate change, which is sometimes called the greenhouse effect, will be a very serious problem for you and your family, somewhat of a problem for you and your family or won't really be a problem for you and your family?

(1b) Do you think that climate change will be a very serious problem for the country as a whole, somewhat of a problem or won't really be a problem for the country as a whole?

(1c) Do you think that climate change will be a very serious problem for other species of plants and animals, somewhat of a problem or won't really be a problem for other species of plants and animals?

(2a) Next, I'd like you to consider the problem of loss of tropical forest. Do you think this will be a very serious problem for you and your family, somewhat of a problem for you and your family or won't really be a problem for you and your family?

(2b) Do you think that loss of tropical forest will be a very serious problem for the country as a whole, somewhat of a problem or won't really be a problem for the country as a whole?

(2c) Do you think that loss of tropical forest will be a very serious problem for other species of plants and animals, somewhat of a problem or won't really be a problem for other species of plants and animals?

(3a) Next, I'd like you to consider the problem of toxic substances in air, water and the soil. Do you think that this will be a very serious problem for you and your family, somewhat of a problem for you and your family or won't really be a problem for you and your family?

(3b) Do you think that toxic substances in air, water and the soil will be a very serious problem for the country as a whole, somewhat of a problem or won't really be a problem for the country as a whole?

(3c) Do you think that toxic substances in air, water and the soil will be a very serious problem for other species of plants and animals, somewhat of a problem or won't really be a problem for other species of plants and animals?

Ascription of responsibility or locus of control

The government should take stronger action to clean up toxic substances in the environment

I feel a personal obligation to do whatever I can to prevent climate change

I feel a sense of personal obligation to take action to stop the disposal of toxic substances in the air, water, and soil

Business and industry should reduce their emissions to help prevent climate change

The government should exert pressure internationally to preserve the tropical forest

The government should take strong action to reduce emissions and prevent global climate change

Companies that import products from the tropics have a responsibility to prevent destruction on the forests in those countries

People like me should do whatever we can to prevent the loss of tropical forests

The chemical industry should clean up the toxic waste products it has emitted into the environment

Figure 3.1 depicts the model proposed to explain sustainable behavior. From the left side of the figure, situational factors (demographics, in this study) which either counteract or strengthen actions in the model are taken into account. In order to discern a personally or socially preferable way of life, the four core values based on

Table 3.3 Emotional competencies selected by the author and psychological dimensions (based on Boyatzis et al. 2002)

Category	Emotional competence	Associated psychological dimension
Self-knowledge		
Confidence in oneself	Believes one-self to be capable for a job Doubts his/her own ability	Effectiveness 1
Self-management		
Emotional self-control	Acts impulsively Stays composed and positive, even in stressful situations	Anticipation 1
Integrity	Keeps his/her promises Acknowledges mistakes	Austerity 1
Adaptability	Adapts ideas based on new information Changes overall strategy, goals, or projects to fit the situation	Austerity 2
Orientation to achievement	Anticipates obstacles to a goal Takes calculated risks to reach a goal	Effectiveness 2
Initiative	Hesitates to act on opportunities Cuts through red tape or bends rules when necessary	Deliberation 1
Understanding of other		
Empathy	Relates well to people of diverse backgrounds Can see things from someone else's perspective?	Solidarity 1
Social skills		
Leadership	Leads by example Articulates a compelling vision	Deliberation 2
Catalyst for change	Personally leads change initiatives Advocates change despite opposition	Anticipation 2
Teamwork	Solicits others' input Establishes and maintains close relationships at work	Solidarity 2

Effectiveness is the tendency to respond swiftly to demands
Deliberation is the act of directing actions toward a specified end
Anticipation is the expectation of future actions or outcomes
Solidarity is the tendency to be concerned about and to act in favor of others
Austerity is prudent and conservative behavior in the face of an uncertain world

inherent conflicts or compatibility among people's motivational goals are taken into account. Two personality traits (ascription of responsibility and awareness of consequences) inform us as to people's desire to take action on the environmental issues. Two key elements of personal skills—inter- and intrapersonal intelligences—which are concerned with the capacity to understand the intentions, motivations, and desires of others and oneself, are considered. These two personal skills were analyzed through the five psychological dimensions to predict sustainability actions of

HE subjects. The author of this study believes that both the psychological and the demographic variables elucidate people's sustainable behavior. That is, human sustainable behavior is based on core elements of personality which determine an action in favor of the common good, as well as causal factors joined to both the idea of sustainable actions and to social and individual responsibility in any culture.

3.3 Research Design

This section discusses the relation of the theoretical approach with the research design. Chapter 1 stated that UNESCO has promoted the education for sustainable development and has emphasized the importance of sustainability in educational curricula from nursery to higher education. Higher educational institutions world-wide made significance efforts in establishing agreements to face the sustainable development challenges before the decade of education for sustainable development was issued, as we can see in different international summits.

In 1990, 318 HEI participated in the Presidents' Conference in France and signed the Talloires Declaration (IISD 1996e; AULSF 2002), stating that environmental changes threaten the survival of humans and thousands of other species, the earth's integrity and biodiversity, the security of nations, and the heritage of future generations.

In 1991, another 33 universities of 10 countries in all continents attended and signed the declaration of the Sustainable Development University Action Conference in Halifax, Canada (IISD 1996b). The Halifax Declaration expressed dismay regarding continuing widespread degradation of the earth's environment and the pervasive influence of poverty on such environmental degradation as well as current widespread unsustainable environmental practices.

In 1991, an initial 29 universities, and two years later another 213 universities, signed the Copernicus Charter at the European Rectors Conference in Barcelona, Spain (IISD 1996a). This charter expresses a collective commitment and represents an effort to mobilize the resources of higher education institutions in order to clarify the concept and further sustainable development objectives.

In 1993, four hundred universities of 47 countries in the Association of Commonwealth Universities attended the Fifteenth Quinquennial Conference in Swansea, Wales (IISD 1996d). Focusing on the topic of people and the environment, they sought ways these universities could respond appropriately to the environmental challenge.

In November 1993, the International Association of Universities, in its 8th Round Table meeting in Japan, issued a clarion call to its 650 university members on the topic of sustainability in the Kyoto Declaration (IISD 1996c).

In October 2001, the International COPERNICUS Conference on Higher Education for Sustainability, organized by the European University Association, took place at the University of Lüneburg in Germany. Participants adopted the Lüneburg Declaration (AULSF 2002) which calls for HEI, NGO's, governments, and United Nations Agencies to support and ensure the introduction of sustainability in their programs and research.

HEI are important actors however who implement actions on teaching, research, outreach, and campus management in HEI are university authorities, faculty, administrators, and students. Therefore, these people provide a vision of what can be sustainable behavior. All items presented in Appendix C refer to the perception of these people. Sustainable behavior is measured on an individual basis; however, it is estimated by statistics as a collective attitude.

3.3.1 The Selection of Participants

The research is conducted taking into account the construct (theoretical concept) of SB and four latent variables (hypothetical terms) in four universities, for three reasons: First, to test the SB perception in universities with vastly different cultures and economies; second, to test if any difference exists among university people and outsiders. The third reason relates to extend the knowledge and experience of the author about her own university on SB perception.

To test the hypotheses, data were collected through either pen-and-paper or Web-nested questionnaire to students, administrators, faculty, and authorities of a Mexican, a German, and three French-speaking Swiss and Canadians universities. Outsiders were all Mexicans who were not related to any HEI.

The selection of participants was limited to those located in:

1. The *Universidad Autónoma Metropolitana, Azcapotzalco* (UAMA), which is located north of Mexico City, is one of five campuses of the UAM, a public university. In 2006, the UAM issued a general framework, the so-called *Plan Institucional hacia la Sustentabilidad*. This plan was part of a broader program developed by a three-part initiative of the Mexican Environmental Ministry, the National Association of Universities and Higher Education Institutions, and the Center for University Research. This initiative was published in 2000 and encourages a strategy to lead HEI toward improved environmental performance in light of the Decade of Education for Sustainable Development (Juarez-Nájera et al. 2006a).

2. The *Leuphana Universität Lüneburg, Institut für Umweltkommunication* (LULIfUK), a public university 30 km from Hamburg in the Federal Republic of Germany, honored with the UNESCO Chair in Higher Education for Sustainable Development. The central aim of the UNESCO Chair is to investigate how academic teaching and learning can be reoriented toward sustainable development.

3. The *Université de Genève* (UdeG), *Institute of Economics and Econometrics*, a public university located in the city of Geneva, Switzerland's second largest university. UdeG enjoys an international reputation which has been won due to its strong ties to many international Geneva-based organizations, such as the

World Health Organization, the International Telecommunications Union, the International Committee of the Red Cross, and the European Organization for Nuclear Research.
4. The *Université de Montréal* (UdeM) and the *Université de Québec à Montréal* (UQAM), *Master Program of Museology* (co partnered by both universities in their Faculties of art and sciences), are public francophone universities located in Montreal, Province of Quebec. UdeM is the second largest in Canada and UQAM the third one in students enrolled.

3.3.2 Data Collection Procedure

The application of the questionnaire was done in two stages. The first stage was carried out in 2008. Only two samples were obtained: The UAMA questionnaire was applied directly to 69 participants who were key individuals; that is, they are or have been members of one of the three campus councils or have coordinated activities providing support and service for the entire campus community. At LULIfUK, the questionnaire was applied through the Internet via 37 participants' e-mails. Each participant's decision-making activities are unknown. At this stage, it was possible to explore statistically the structure of SB construct as a priori model (Juárez-Nájera 2010; Juárez-Nájera et al. 2010).

The second stage was carried out in 2013 using a larger number of participants (in total 218) adding 95 Mexican outsiders. The UAMA questionnaire was applied directly to 58 participants either by pen-and-paper questionnaire or by Internet in the same Web-nested questionnaire. At LULIfUK, UdeG, UdeM, and UQAM, the web-nested questionnaire was applied to 40 German participants, 19 Swiss participants, and 9 Canadian participants.

For testing the hypothesis, four samples of participants were prepared: the so-called All-HEI sample (232) which includes all university participants at both stages and from the five universities. The UAMA sample (127 participants) was called Lower Socio-Economic Level University (LSELU), and a No-UAMA sample (105 participants) was called Higher Socio-Economic Level Universities (HSELU). This latter was without outsiders (95 participants), only university participants from richer countries (Germany, 40 participants; Switzerland, 19 participants; and Canada, 9 participants).

3.3.3 Questionnaire Design

The design of an open questionnaire facilitates data collection and statistical data handling when time is pressing the outputs as was the case in the first stage of this study. Regarding the number of items (Aiken 1997; McDonald 2011) that make up a questionnaire, Montalvo (2002) mentions that it is accepted that people's capacity for information processing in decision problems is limited. In this study, an open questionnaire was prepared which consisted of 67 items in five sections according to the four latent variables and demographics. The questionnaire is included in Appendix C.

The first section of universal values included 21 items of Schwartz's (1994), Schwartz and Boehnke (2004) 10 value categories. At least one item was included from each value type. Fifteen of the items supported principles underlying the ESD (items 1.1, 1.4, 1.6, 1.8, 1.9, 1.11, 1.13 to 1.21) and six items were contrary to ESD (items 1.2, 1.3, 1.5, 1.7, 1.10 and 1.12). The order of all variables was randomized to prevent participants from anticipated response and their scales were inverted for statistical treatment.

The variables for moral norm activation from the second and third sections of the questionnaire were measured through nine items regarding awareness of consequences (AC) and nine regarding ascription of responsibility (AR). Those questions related to AC included importance to oneself, country, and other species of three actual environmental problems (climate change, loss of forests, and chemicals). In the AR section, three items concerned personal obligations, three concerned government obligations, and three concerned business obligations (Stern et al. 1999).

The fourth section on intrapersonal and interpersonal intelligences contained 20 items from Boyatzis et al. (2002), analyzed through five psychological dimensions of sustainability from Corral-Verdugo (2010) and Corral-Verdugo and Pinheiro (2004). The order of these variables was randomized to prevent participants from anticipated response.

The final section contained eight questions related to demographics such as age, gender, religious denomination, general income level, and educational training (Hines et al. 1987). These variables were dichotomous. Fifty-nine items were polytomous in four different Likert scale items: Thirty items corresponded to a multiple choice among fully agree, agree, undecided, disagree, and strongly disagree. Nine items were answered with very serious, somewhat serious, or not serious. Twenty items were answered with never, rarely, sometimes, many times, and constantly (Converse and Presser 1988; Kirakowski 2000).

Polytomous models show the relationship of a variable of a latent trait variable in an ongoing way (Henerson et al. 1991; Ligtvoet et al. 2011; Shiken 2000) and are used usually because they are more informative and reliable than items with dichotomous scores (Embretson and Reise 2000). However, Scheuthle et al. (2005) indicate that, contrary to common expectations, a broader set of questions causes more diverse, arbitrary participant responses.

Stern et al.'s (1999) questionnaire, as well as the personal intelligence competence list, was translated from English to Spanish. Subsequently natives speaking German and French translated the Spanish questionnaire to German and French, and then these were reviewed by LULIfUK, UdeG, UdeM, and UQaM staff. Spanish, German, and French questionnaires are not included in this manuscript but are available upon request to the author.

In the first stage, the questionnaire was prepared using the *Pinpoint* software version 3.10 (1995) which facilitated the task of capturing individual's responses. Thus, basic descriptive statistics were run and handled subsequently into SAV format by using SPSS software version 12, and then a principal component analysis (Jollife 1998) was carried out. In order to determine item response probability calculations, Microsoft Excel format was used.

In the second stage, the Web-nested questionnaire was prepared in a free platform in the four languages mentioned above; it is hosted in UAMA Web site. Subsequently input matrixes were run into SAV format by using SPSS (2013) software (version 20: Arbuckle 2012), and then a principal component analysis were carried out (Barendse et al. 2014; Brown 2006; Schmitt 2011). After that, using the *Mplus* software (version 6: Muthén and Muthén 1998–2010), a confirmatory factor analysis was carried out (Asparouhov and Muthén 2014; Brown 2006; Finch and Bronk 2011).

The limitations of the study are related to methodological approach adopted. This is related to the inferences that can be reasonably made about relations among the SB construct and latent variables when all such variables are assessed with open questionnaires. The application of the norm-activation framework, extended by Stern et al. (1999) and Hines et al. (1987) models, assumes that people behave according to their own values, their personality traits, and their skills supported by situational factors. That is, personal norms leading to sustainable actions are activated by individual's belief that several conditions threaten things which they value, and that they can act to reduce the threat. These norms create a general predisposition which affects many types of behaviors carried out with sustainable intentions. The statistical analysis carried out tests the reliability and degree of association between these variables; some statistical estimators are available upon request to the author. In the following Chap. 4, a description of the specific factors of sustainable behavior will be presented.

References

Aiken, L. R. (1997). *Questionnaires and inventories: Surveying opinions and assessing personalities.* NYC: John Wiley and Sons.

Arbuckle, J. L. (2012). *User's guide: IBM-SPSS Amos 21.* Amos Development Corp.

Asparouhov, T., & Muthén, B. (2014). Multiple-group factor analysis alignment. *Structural Equation Modeling: A Multidisciplinary Journal 21*, 1–14. doi:10.1080/10705511.2014.919210.

AULSF—Association of University Leaders for a Sustainable Future. (2002). Talloires declaration, Resource kit. Washington: AULSF.

Barendse, M. T., Oort F. J., & Timmerman, M. E. (2014). Using exploratory factor analysis to determine the dimensionality of discrete responses. *Structural Equation Modeling: A Multidisciplinary Journal.* doi:10.1080/10705511.2014.934850.

Boyatzis, Goleman, Hay Acquisition Co. Inc. (2002). *Inventario de competencias emocionales.* HayGroup.

Brown, T. A. (2006). *Confirmatory factor analysis for applied research.* NYC: The Guilford Press.

Converse, J. M., & Presser, S. (1988). *Survey questions. Handcrafting the standardized questionnaire* (2nd Printing). University Press series no. 07-063. Beverly Hills: Sage Publications.

Corral-Verdugo, V. (2010). *Psicología de la sustentabilidad: un análisis que nos hace pro ecológicos y pro sociales.* México: Trillas.

Corral-Verdugo, V., & Pinheiro, J. (2004). Aproximaciones al estudio de la conducta sustentable. *Medio Ambiente y Comportamiento Humano, 5*(1 y 2), 1–26.

Dunlap, R. E., Van Liere, K. D, Mertig, A. G., & Jones, R. E. (2000). Measuring endorsement of the new environmental paradigm. A revised NEP scale. *Journal of Social Issues, 56,* 3, 425–442.

Dunlap, R. E., & van Liere, K. D. (1978). The new environmental paradigm. *Journal of Environmental Education, 9,* 10–19.

Embretson, S. E., & Reise, S. P. (2000). *Item response theory for psychologist.* Mahwah, New Jersey: Lawrence Erlbaum Associates, Publishers.

Finch, W. H., & Bronk, K. C. (2011). Conducting confirmatory latent class analysis using Mplus. *Structural Equation Modeling: A Multidisciplinary Journal, 18*(1), 132–151. doi:10.1080/10705511.2011.532732.

Hawcroft, L. J., & Milfont, T. L. (2010). The use (and abuse) of the new environmental paradigm scale over the last 30 years. A *Meta-analysis. Journal of Environmental Psychology, 30,* 143–158.

Henerson, M. E., Morris, L. L., & Fitz-Gibbon, C. T. (1991). *How to measure attitudes* (2nd ed). NYC: Sage Publication/The Regents of the University of California (1987).

Hines, J. M., Hungerford, H. R., & Tomera, A. N. (1987). Analysis and synthesis of research on responsible environmental behavior: A meta-analysis. *Journal of Environmental Education, 18* (2), 1–8.

IISD—International Institute for Sustainable Development. (1996a). Copernicus charter. http://www.iisd.org/educate/declare.htm. (December, 2006).

IISD—International Institute for Sustainable Development. (1996b). Halifax declaration. http://www.iisd.org/educate/declarat/halifax.htm. (December, 2006).

IISD—International Institute for Sustainable Development. (1996c). Kyoto declaration. http://www.iisd.org/educate/declare.htm. (December, 2006).

IISD—International Institute for Sustainable Development. (1996d). Swansea Declaration. http://www.iisd.org/educate/declare.htm (December, 2006).

IISD—International Institute for Sustainable Development. (1996e). Talloires Declaration. The Role of Universities and University Presidents in Environmental Management and Sustainable Development. In: *Actual signatory list Report and Declaration of the Presidents' Conference, Tufts University European Center, Talloires, France* (October 4–7, 1990). http://www.iisd.org/educate/declare.htm. (December, 2006).

Juárez-Nájera, M. (2010). *Sustainability in higher education. An explorative approach on sustainable behavior in two universities.* Doctoral Thesis. Faculty of Social Sciences, Erasmus University Rotterdam, The Netherlands.

Juárez-Nájera, M., Rivera-Martínez, J. G., & Hafkamp, W. A. (2010). An explorative socio-psychological model for determining sustainable behavior: Pilot study in German and Mexican Universities. *Journal of Cleaner Production, 18,* 686–694.

Juárez-Nájera, M., Dieleman H., & Turpin-Marion, S. (2006a). Sustainability in Mexican Higher Education, towards a new academic and professional culture. *Journal of Cleaner Production.* Special issue on education for sustainable development. *14*(9–11), 1028–1038.

Jolliffe, I. T. (1986). Principal component analysis. In D. Brillinger, S. Fienberg, J. Gani, J. Hartigan, & K. Krickeberg. (Advisors) *Springer Series in Statistics.* New York: Springer.

Kaiser, F. G., Hübner, G., & Bogner, F. (2005). Contrasting the theory of planned behavior with the value-belief-norm model in explaining conservation behavior. *Journal of Applied Social Psychology, 35*(10), 2150–2170.

Kirakowski, J. (Comp.) (2000). *Questionnaires in usability engineering. A list of frequently asked questions.* Cork, Ireland: Human Factors Research Group.

Ligtvoet, R., Van der Ark, L. A., Bergsma, W. P., & Sijtsma, K. (2011). Polytomous latent scales for the investigation of the ordering of items. *Psychometrika, 76*(2), 200–216.

McDonald, R. P. (2011). Measuring latent quantities. *Psychometrika, 76*(4), 511–536.

Montalvo, C. (2002). *Environmental policy and technological innovation, why do firms adopt or reject new technologies?* Cheltenham, UK: Edward Elgar.

Muthén, L. K., & Muthén, B. O. (1998–2010). *Mplus user's guide* (6th ed.). Los Angeles, CA: Muthén & Muthén.

PinPoint. (1993–1995). PinPoint, single user, version 3.10a. UK: Published by Longman Logotron.

Scheuthle, H., Carabias-Hütter, V., & Kaiser, F. G. (2005). The motivational and instantaneous behavior effects of contexts: Step toward a theory of goal-directed behavior. *Journal of Applied Social Psychology, 35*(10), 2076–2093.

Schmitt, T. A. (2011). Current methodological considerations in exploratory and confirmatory factor. *Journal of Psychoeducational Assessment, 2011*(29), 304. doi:10.1177/0734282911406653.

Schwartz, S. H. (1977). Normative Influences on Altruism. In L. Berkowitz (Ed.), *Advances in experimental social psychology* (Vol. 10, pp. 221–279). New York: Academic Press.

Schwartz, S. H. (1994). Are there universal aspects in the structure and contents of human values? *Journal of Social Issues, 50*, 19–45.

Schwartz, S. H., & Huismans, S. (1995). Value priorities and religiosity in four western religions. *Social Psychology Quarterly, 58*, 88–107.

Schwartz, S. H., & Boehnke, K. (2004). Evaluating the structure of human values with confirmatory factor analysis. *Journal of Research in Personality, 38*, 230–255.

Shiken, (2000). What issues affect Likert-scale questionnaire formats? *JALT Testing & Evaluation SIG Newsletter, 4*, 18–21.

SPSS. (2013). *Statistical package for the social sciences*, version 21.0 for Windows. SPSS Inc.

SPSS. (1989–2003). *Statistical package for the social sciences*, version 12.0 for Windows. SPSS Inc.

Stern, P. C., Dietz, T., Abel, T., Guagnano, G. A., & Kalof, L. (1999). A value-belief-norm theory of support for social movements: The case of environmentalism. *Human Ecology Review, 6*, 81–97.

Chapter 4
Mapping Latent Variables of Sustainable Behavior

In the previous chapter, the author proposed a model to describe sustainable behavior (see Fig. 3.1) and explained how to built a questionnaire to test such proposed model among HEIs in different countries. In this chapter, firstly, an exploratory factor analysis (EFA) was used to prespecify the factors in terms of how well it reproduces the sample correlation (covariance) matrix of the measured variables to establish the underlying structure of SB construct. Secondly, a confirmatory factor analysis (CFA) was used to guide the specification and evaluation of the factor model based on a strong empirical, conceptual foundation to validate such SB construct.

4.1 Searching Structure in Multivariate Statistical Procedures

The SB construct and four latent variables (universal values, awareness of consequences, ascription of responsibility, and inter- and intrapersonal intelligences associated with psychological dimension in the proposed SB model of Fig. 3.1) are entities which are impossible to directly observe and measure. The same is true for notions such as "quality of life," "general intelligence," "business sentiment," or "human nature," to name a few. In order to resolve this limitation, social psychologists and other social scientists have theorized and proposed latent variables. Latent variables are mental constructs which represent complex relationships; when subjects respond to a questionnaire containing a variety of indicators, these latent variables may be measured as real entities (Bartholomew 1987; Bollen 2002; McDonald 2011).

The search for structure in correlated psychological variables has been one of the main objectives in psychometrics (e.g., evaluation of multiple-item testing instruments) for several decades. Jöreskog (1978) states that traditionally this search was done by using factor analysis to detect and assess latent sources of variation among a set of observed measures. Since its inception a century ago by Spearman in 1904, factor analysis has become one of the most widely used multivariate statistical procedures in applied research endeavors across a multitude of domains

© Springer International Publishing Switzerland 2015

M. Juárez-Nájera, *Exploring Sustainable Behavior Structure in Higher Education*,
Management and Industrial Engineering, DOI 10.1007/978-3-319-19393-9_4

(educational, organizational, cross-cultural, personality, and social-psychology). The fundamental intent of factor analysis is to determine the number and nature of latent variables (factors) that influences more than one observed measures and that accounts for the correlations among these observed measures (Bollen and Lenox 1991; Bollen 2002, 2007; Brown 2006).

Thurstone's common factor model postulates that each indicator in a set of observed measures is a linear function of one or more common factors and one unique factor. Thus, factor analysis partitions the variance[1] of each indicator into two parts: (1) the variance accounted for by the latent factor and (2) a combination of reliable variance that is specific to the indicator and random error variance (Jöreskog 1978). There are two main types of analyses based on the common factor model: exploratory factor analysis (EFA) and confirmatory factor analysis (CFA) (Brown 2006).

The researcher employs EFA as descriptive technique to determine the appropriate number of common factors and to uncover which measured variables are reasonable indicators of the various latent dimensions. In CFA, the researcher specifies the number of factors and the pattern of indicator–factor loadings in advance, as well as other parameters such as those bearing on the independence or covariance of factors and indicator unique variances (Brown 2006).

4.1.1 Exploratory Factor Analysis

Although related to EFA, principal component analysis (PCA) is frequently miscategorized as an estimation method of common factor analysis. However, PCA relies on a different set of quantitative methods that are not based on the common factor model. PCA does not differentiate common and unique variance; it does not test hypotheses by means of a formal test of significance. Rather, PCA aims to account for the variance in the observed measures rather than explain the correlations among them; it explores the possibility of a factor structure underlying the variables (Basilevsky 1994; Brown 2006; Gardner 2003; Jolliffe 1986). Therefore, PCA provides a large quantity of information, which the researcher can then use to specify factors in future studies.

PCA is more appropriately used as a data reduction technique to diminish a larger set of measures to a smaller, more manageable number of composite variables to use in subsequent analyses. PCA conserves as much of the actual variation in the entire set of data as possible. This reduction of variables is achieved by transforming the data into a new set of variables, principal components, which are not correlated and are ranked according to the first few variables which maintain the majority of the variation present in all original variables (Jolliffe 1986). One use of

---

[1]Variance (σ^2): The measure of variability produced by tacking the average of the sum of the squared deviation from the mean (Welkowitz et al. 2002).

principal component analysis is to establish one or more factors which underlie a large number of variables. As a result, the analysis identifies the number of factors and which variables make up which factor (Gardner 2003; Brace et al. 2006).

Some methodologists (Brown 2006) have argued that PCA is a reasonable superior alternative to EFA, in view of the fact that PCA possesses several desirable statistical properties: computationally simpler, not susceptible to improper solutions, often produces results similar to those of EFA, ability of PCA to calculate a participant's score on a principal component, whereas the indeterminate nature of EFA complicates such computations, calculations are direct and apparently simple and have a wide variety of applications. Specifically, PCA calculation reduces the solution of the eigenvalue (own values) problem to eigenvectors by using a symmetric, semi-defined, positive matrix (Jolliffe 1986).

4.1.2 Confirmatory Factor Analysis

Brown (2006) states that CFA is a type of structural equation modeling that deals specifically with measurement models, that is, the relationships between observed measures (indicators) and latent variables (factors). A fundamental feature of CFA is its hypothesis-driven nature (Dunn et al. 2014; Howard 2013; Ligtvoet et al. 2011). CFA has become one of the most commonly used statistical procedures in applied research. This is because CFA is well equipped to address the types of questions that researchers often ask. Some of the most common uses of CFA are psychometric evaluation of test instrument (to verify the number of underlying dimension of the instrument and the pattern of item–factor relationships), construct validation (the results of CFA can provide compelling evidence of the convergent and discriminant validity of theoretical constructs), method effects (some of the covariation of observed measures is due to sources other than the substantive latent factors), and measurement invariance evaluation (ability to determine how well measurement models generalize across groups of individuals or across time) (Brown 2006).

The author of this study also found that CFA can address the relationships showed in Fig. 3.1 where the proposed SB model depicts participants' willingness in HEI to perform SB which could be explained primarily in terms of attitudes toward the behavior. That is, to validate a second-order SB construct, or to show evidence in the latent structure of four factors, or to explain the effect of categorical indicators.

4.2 Results of a Principal Component Analysis

The author of this research prepared a multi-item questionnaire of sustainable behavior. The questionnaire was administered to 327 adults, a modest sample. PCA was conducted with *SPSS* (version 20: for further discussion see Arbuckle 2012).

PCA statistical technique may determine the appropriate number of common factors and uncover which measured variables are reasonable indicators of the various latent dimensions (Barendse et al. 2014; Brown 2006; Schmitt 2011).

PCA (for more details on calculations, see Appendix D.1) results appear according to the following procedures: (1) identification of the number of factors to foster the interpretability of the solution; and (2) estimation of factor scores which show representative relations of the latent variables. Many procedural statistics are not shown due to the concise outputs in this research; however, they are available upon request to the author as indicated.

4.2.1 Identification of the Number of Factors

This section explains the descriptive statistics to show minimum, maximum, means, and variances of participants for 61 indicators, the original matrix of data set, factorability indicators as Kaiser–Meyer–Olkin and Bartlett's test, the anti-image matrix, communalities to indicate how much variance within each variable is explained by the analysis, the scree plot, eigenvalues to explain variance of sample, the component matrix, the reproduced matrix, and the matrix of rotated factors.

4.2.1.1 Descriptive Statistics of Four Samples

Table 4.1 shows minimum and maximum values as well as means and variances of categorical responses (mentioned in Chap. 3, questionnaire design section, and in Appendix C) for a total of 313 participants in 59 indicators (or items) and two demographics within a questionnaire. The first and third sections of the questionnaire have minimum and maximum values which range from 1 to 5: that is, minimum and maximum participant's responses totally agree to totally disagree. The second section ranges from 1 to 3: that is, minimum and maximum participant's responses very serious to no serious. The fourth section ranges from 1 to 5; that is, minimum and maximum participant's responses never, rarely, sometimes, often, and consistently achieve. The fifth section for demographics has dichotomous variables with yes or no responses.

Table 4.1 is also grouped in four samples: Lower Socio-Economic Level University (LSELU) which is the Mexican university with 127 participants, a No-LSELU sample which was called Higher Socio-Economic Level Universities (HSELU) including participants ($N = 105$) from richer-country (Germany, Switzerland, and Canada) universities. Also it was prepared an All-HEI sample including responses from all university participants ($N = 232$), and an Outsiders sample including responses from Mexicans ($N = 95$) who are not members of the Mexican university. Regarding the comparison of results, two groups were associated: (1) LSELU versus HSELU, and (2) All-HEI versus Outsiders.

Table 4.1 Descriptive statistics of four samples

Latent variable/item or indicator	LSELU sample N = 127				HSELU sample N = 105				All-HEI sample N = 232				Outsiders sample N = 95			
	Min.	Max.	Mean	Variance	Min.	Max.	Mean	Variance	Min.	Max.	Mean	Variance	Min.	Max.	Mean	Variance
I. Universal values																
1 Worldatpeace	1	5	1.55	1.075	1	5	1.44	0.710	1	5	1.50	0.909	1	5	1.78	1.451
2 Influential	1	5	3.99	0.833	1	5	3.41	0.783	1	5	3.73	0.891	2	5	3.82	0.851
3 Ambitious	1	5	2.73	1.309	1	5	2.66	0.901	1	5	2.70	1.121	1	5	3.09	1.108
4 Broadminded	1	5	1.39	0.558	1	5	1.50	0.599	1	5	1.44	0.577	1	4	1.66	0.822
5 Authority	1	5	2.98	1.515	1	4	2.40	0.915	1	5	2.72	1.321	1	5	3.29	1.104
6 Creativity	1	5	1.27	0.404	1	4	1.59	0.590	1	5	1.41	0.512	1	5	1.54	0.741
7 Socialpower	1	5	2.79	1.296	1	4	1.78	0.673	1	5	2.33	1.262	1	5	2.97	1.286
8 Socialorder	1	4	1.64	0.582	1	5	2.36	0.772	1	4	1.97	0.795	1	5	2.05	1.050
9 Prevention	1	4	1.17	0.250	1	5	1.26	0.424	1	4	1.21	0.329	1	4	1.38	0.642
10 Variedlife	1	5	4.61	0.429	1	5	4.16	0.560	1	5	4.41	0.537	1	5	4.35	0.655
11 Socialjustice	1	4	2.23	0.844	1	5	1.52	0.617	1	4	1.91	0.862	1	5	2.13	0.856
12 Enjoyinglife	2	5	4.72	0.379	1	5	4.46	0.481	2	5	4.60	0.440	1	5	4.42	0.693
13 Selfdiscipline	1	4	1.87	0.688	1	5	2.44	0.806	1	4	2.13	0.820	1	5	1.75	0.893
14 Unitywnature	1	4	1.48	0.490	1	4	1.80	0.738	1	4	1.62	0.625	1	5	1.48	0.635
15 Health	1	4	1.20	0.339	1	5	1.49	0.656	1	4	1.33	0.500	1	5	1.45	0.697
16 Responsible	1	5	1.28	0.490	1	5	1.57	0.632	1	5	1.41	0.573	1	5	1.40	0.604
17 Respectful	1	4	1.22	0.284	1	4	1.48	0.502	1	4	1.34	0.397	1	5	1.37	0.725
18 Moderate	1	4	1.79	0.708	1	5	2.41	0.821	1	4	2.07	0.852	1	5	1.91	0.704
19 Equality	1	4	1.39	0.447	1	5	1.50	0.522	1	4	1.44	0.482	1	5	1.53	0.763
20 Acceptinglife	1	5	2.75	1.936	1	5	3.38	1.257	1	5	3.03	1.722	1	5	2.72	1.865
21 Choosinggoals	1	4	1.37	0.441	1	5	1.62	0.546	1	4	1.48	0.502	1	4	1.53	0.443

(continued)

Table 4.1 (continued)

Latent variable/item or indicator	LSELU sample N = 127				HSELU sample N = 105				All-HEI sample N = 232				Outsiders sample N = 95			
	Min.	Max.	Mean	Variance	Min.	Max.	Mean	Variance	Min.	Max.	Mean	Variance	Min.	Max.	Mean	Variance
II. Awareness of consequences																
22aClimateyou	1	3	1.24	0.234	1	3	1.76	0.452	1	3	1.48	0.398	1	3	1.33	0.307
23bClimatecountry	1	3	1.19	0.186	1	3	1.45	0.327	1	3	1.31	0.265	1	3	1.21	0.189
24cClimateplants	1	3	1.15	0.160	1	3	1.15	0.169	1	3	1.15	0.163	1	2	1.15	0.127
25aForestyou	1	3	1.26	0.226	1	3	1.77	0.486	1	3	1.49	0.407	1	3	1.21	0.211
26bForestcountry	1	3	1.17	0.187	1	3	1.51	0.387	1	3	1.32	0.306	1	2	1.12	0.103
27cForestplants	1	3	1.08	0.089	1	2	1.06	0.054	1	3	1.07	0.073	1	2	1.09	0.087
28aToxicyou	1	2	1.10	0.093	1	3	1.40	0.319	1	3	1.24	0.216	1	3	1.16	0.156
29bToxiccountry	1	2	1.10	0.093	1	3	1.33	0.263	1	3	1.21	0.182	1	3	1.16	0.156
30cToxicplants	1	2	1.09	0.086	1	2	1.14	0.124	1	2	1.12	0.103	1	2	1.15	0.127
III. Ascription of responsibility																
31Government1	1	5	1.32	0.411	1	4	1.60	0.646	1	5	1.45	0.534	1	5	1.33	0.541
32Ifeelobligation	1	5	1.46	0.504	1	5	1.79	0.648	1	5	1.61	0.594	1	3	1.53	0.443
33Ifeelasense	1	3	1.52	0.394	1	5	1.89	0.775	1	5	1.69	0.598	1	4	1.47	0.465
34Business	1	5	1.25	0.380	1	3	1.19	0.233	1	5	1.22	0.313	1	4	1.22	0.323
35Government2	1	3	1.21	0.232	1	4	1.45	0.480	1	4	1.32	0.357	1	3	1.19	0.240
36Government3	1	5	1.35	0.485	1	3	1.31	0.275	1	5	1.34	0.389	1	4	1.29	0.423
37Companies	1	5	1.35	0.516	1	4	1.70	0.849	1	5	1.51	0.693	1	4	1.28	0.397
38Peoplelike	1	5	1.52	0.474	1	5	1.98	0.884	1	5	1.73	0.709	1	4	1.46	0.507
39Industry	1	3	1.25	0.222	1	3	1.28	0.317	1	3	1.26	0.264	1	3	1.15	0.170

(continued)

Table 4.1 (continued)

Latent variable/item or indicator	LSELU sample N = 127				HSELU sample N = 105				All-HEI sample N = 232				Outsiders sample N = 95			
	Min.	Max.	Mean	Variance	Min.	Max.	Mean	Variance	Min.	Max.	Mean	Variance	Min.	Max.	Mean	Variance
IV. Personal intelligences																
40Anticipatesobsta	2	5	3.80	0.909	1	5	3.39	0.753	1	5	3.62	0.748	2	5	3.27	0.839
41Adaptsideas	2	5	3.93	0.818	2	5	3.71	0.646	2	5	3.83	0.565	1	5	3.44	0.739
42Solicitsinput	1	5	3.94	0.974	1	5	3.35	0.796	1	5	3.67	0.888	1	5	3.33	0.882
43Takesrisks	1	5	3.76	0.861	2	5	3.20	0.789	1	5	3.50	0.762	1	5	3.13	0.601
44Relateswell	2	5	4.28	0.861	2	5	4.22	0.665	2	5	4.25	0.604	1	5	3.89	0.904
45Stayscomposed	1	5	3.85	0.892	2	5	3.60	0.816	1	5	3.74	0.749	1	5	3.24	1.037
46Leadsbyexam	2	5	4.02	0.816	2	5	3.19	0.637	2	5	3.64	0.716	1	5	3.52	0.869
47Advocateschange	2	5	4.02	0.840	2	5	3.47	0.809	2	5	3.77	0.757	1	5	3.42	0.970
48Actsimpulsively	1	5	2.75	0.845	1	5	2.95	0.764	1	5	2.84	0.663	1	5	2.88	0.891
49Personallyleads	1	5	3.68	0.950	1	5	2.69	0.974	1	5	3.23	1.164	1	5	2.87	0.877
50Keepspromi	2	5	4.41	0.634	3	5	4.22	0.604	2	5	4.32	0.393	1	5	3.81	0.815
51Acknowledgeesmist	2	5	4.10	0.744	2	5	3.95	0.656	2	5	4.03	0.501	1	5	3.64	0.871
52Articulatesacompe	2	5	3.90	0.722	1	5	3.27	0.724	1	5	3.61	0.619	1	5	3.41	0.798
53Canseethings	1	5	3.82	0.706	2	5	3.93	0.669	1	5	3.87	0.477	1	5	3.26	1.004
54Believescapable	2	5	4.61	0.578	2	5	3.90	0.687	2	5	4.29	0.520	1	5	4.41	0.585
55Bendsrules	1	5	2.20	0.962	1	4	2.63	0.858	1	5	2.40	0.881	1	4	2.13	0.792
56Doubtsownability	1	5	2.02	1.027	2	5	3.02	0.772	1	5	2.47	1.090	1	5	1.85	1.106
57Establishesclose	1	5	3.98	0.976	2	5	3.80	0.825	1	5	3.90	0.834	1	5	3.69	0.768
58Hesitatestoact	1	5	2.41	0.971	1	5	2.99	0.766	1	5	2.67	0.862	1	5	2.59	1.053
59Changesstrategy	1	5	3.34	1.025	2	5	3.57	0.770	1	5	3.44	0.854	1	5	3.40	0.987
Demographics																
60Gender	1	2	1.43	0.246	1	2	1.65	0.230	1	2	1.53	0.250	1	2	1.45	0.250
61Age	19	85	41.02	193.539	20	78	31.61	99.260	19	85	30.08	209.954	16	81	42.35	206.272

With regard to the analysis of universal value section (indicators or items 1–21 in Table 4.1), LSELU/HSELU group had item 9Prevention the lowest mean (1.17 and 1.26, respectively) and less variability among participants (0.250 and 0.424, respectively), that is, the mean response is totally agree with this value; while item 12Enjoyinglife had the highest mean (4.72 and 4.46, respectively) which means respondents of both groups were strongly disagree to such item. Indicator 10Variedlife had means around 4 and indicators 2Influential, 3Ambitious, 7Socialpower, 5Authority, and 20Acceptinglife showed scores around 2 and 3, that is, the mean response is that the majority of participants somewhat disagree with these values. It is important to mention that these five indicators (10, 2, 3, 7, and 5) were considered to represent the reverse of the goal of EDS (see Table 3.1) and this seems to provide corroborative evidence for the idea proposed in Chap. 2 that values of achievement, hedonism, and stimulation go against the underlying principles of ESD. Indicators 8Socialorder, 13Selfdiscipline, and 18Moderate had means around 1.5–2 which means respondents are somehow agree with this values. The rest of the indicators (1Worldatpeace, 4Broadminded, 6Creativity, 14Unitywnature, 15Health, 16Responsible, 17Respectful, 19Equality, and 21Choosingoals) had scores ranging 1–1.5, clearly representing the response totally agree. About All-HEI/Outsiders group had also the item 9Prevention the lowest mean (1.21 and 1.38, respectively) and less variability among participants. 12Enjoyinglife had the highest mean (4.60 and 4.42, respectively). The rest of the indicators showed scores very similar to that of the LSELU/HSELU group. According to the four universal value categories from Fig. 2.4, the majority of respondents from both groups were totally agree with self-transcendence value (items: 1, 4, 9, 14, and 16) and totally or somehow disagree with openness to change values (items: 12, 10, and 21). These results are explained by the structure of dynamic relationship among universal values.

Awareness of consequences section 2 (items 22a–30c in Table 4.1) shows mean values of both groups on average of 1 in almost all indicators which means that respondents consider those environmental problems (climate change, lost of tropical forest, and toxic substances) as very serious, with a variance lesser or equal to 1, while most seriously when the problem is for oneself (item 22aClimateyou and 25aForestyou) instead of the country or other species.

Ascription of responsibility section 3 (items 31–39 in Table 4.1) shows mean values in both groups (LSELU/HSELU and All-HEI/Outsiders) on average of 1 or close to 2 in almost all indicators which means that respondents consider strongly agree with the kind of obligation to accept either by themselves or government/businesses, with a variance lesser or equal to 1. Indicator 32Ifeelobligation, 33Ifeelasense, and 38Peoplelike (or internal locus of control) present higher mean values, that is, they are clearly closer to agree than is the case for other variables.

Personal intelligence section 4 (indicators 40–59 in Table 4.1) for LSELU/HSELU group had mean participant's responses ranged from 2.02 to 4.61, with a variance lesser or equal to 1. Six indicators had mean values on average of 4 and 3 in 54Believescapable, 50Keepspromi, 44Relateswell, 51Acknowledgesmist, 46Leadsbyexam, and 47Advocateschange which means that respondents do often and sometimes these actions; item 44Relateswell had one of the highest value (4.28 and 4.22, respectively) between this group. Items 56Doubtsownability, 55Bendsrules, 58Hesitatestoact, and 48Actsimpulsively had mean values around 2 which mean they do rarely these actions. The rest of indicators (40Anticipatesobsta, 41Adaptsideas, 42Solicitsinput, 43Takesrisks, 45Stayscomposed, 49Personallyleads, 52Articulatesacompe, 53Canseethings, 57Establishesclose, and 59Changesstrategy) had mean values between 2 and 3, that is, rarely or sometimes respondents achieve them. About All-HEI/Outsiders group showed scores very similar to that of the LSELU/HSELU group. Indicator 54Believescapable had the highest mean value (4.29 and 4.41, respectively) and item 56Doubtsownability the lowest one (2.14 and 1.89, respectively). According the inter- and intrapersonal intelligence categories associated with five psychological dimensions of sustainability from Table 3.3, the majority of respondents from both groups did very often indicator 54Believe-your-self-to-be-capable-for-a-job which was associated with Effectiveness1, one of the psychological dimensions of sustainability. At the same time, the majority of respondents from both groups did rarely 56Doubt-own-ability which was associated with Effectiveness1 too. These results show in randomized items that Effectiveness1, for easier and more difficult actions, is shown in the structure of personal skills. Regarding other personal intelligences, Austerity1 (items: 50, 51) and Deliberation1 (items: 55, 58) were often achieved by participants.

The last section demographics (items 60 and 61 in Table 4.1) for the LSELU/HSELU group had mean value in item 60Gender of 1.43 and 1.65, respectively, clearly more female respondents in both type of universities, while indicator 64Age ranges from 19/20 to 85/78 years with an average of 41 and 31, respectively. About All-HEI/Outsiders group had value in item 60Gender of 1.53 and 1.45, respectively, clearly more female respondents in both types of samples, while indicator 64Age ranges from 19/16 to 85/81 years with an average of 30 and 42, respectively.

4.2.1.2 Correlation Matrix of Four Samples

This matrix describes bivariate relationships involving all variables. The criterion of 0.3 is normally considered the lower cutoff by which variables are factorable according to Brace et al. (2006); however, in this research, 0.400 was the cutoff in order to obtain higher values which provide more reliable conclusions. The original table is not shown in this section, but the table is available upon request to the author. Correlation matrix values were modestly good for four samples.

4.2.1.3 Kaiser–Meyer–Olkin Measure and Bartlett's Test of Sphericity

These tests provide some information regarding data factorability. KMO is a test of the amount of variance within the data which could be explained by factors. A KMO value of 0.5 is poor; a value closer to 1 is better (Brace et al. 2006).

Table 4.2 KMO and Bartlett's test for LSELU, HSELU, All-HEI, and Outsiders samples

Kaiser–Meyer–Olkin measure of sampling adequacy		LSELU 0.628	HSELU 0.525	All-HEI 0.761	Outsiders 0.546
Bartlett's test of sphericity	Approx. chi-square	4023.263	6212.830	3496.009	4121.710
	Degrees of freedom	1830	1830	1830	1711
	Sig.	0.000	0.000	0.000	0.000

Anti-image correlation of the four samples from the anti-image matrices

Bartlett's test shows that the data have a probability of factorability: if data have $p > 0.05$, the test recommends not to continue; but if data have $p < 0.05$, the test recommends to check other indicators of factorability before proceeding (Brace et al. 2006).

The factorability for the entire LSELU, HSELU, All-HEI, and Outsiders samples is presented in Table 4.2. The values of the table are modestly good, but still close to 0.525–0.76. The KMO test for HSELU and Outsiders samples is poor, close to 0.6. Table 4.2 shows the amount of variance within data that could be explained by factors. Bartlett's test indicates in all cases that data are probably factorable because $p < 0.05$.

The upper matrix contains negative partial covariances and the lower matrix contains negative partial correlations. The on-diagonal values in the anti-image correlation matrix are the KMO values. If any variable has a KMO value less than 0.5, one should consider dropping it from the analysis (Brace et al. 2006). The original table is not shown in this section, but the table is available upon request to the author. The KMO values are modestly good. However, high correlation values were obtained from each latent variable group.

4.2.1.4 Communalities of the Four Samples

Communalities indicate how much variance within each indicator or item is explained by the analysis. The initial communalities are calculated using all possible components, and these are always = 1 for each sample. The extraction communalities are calculated using the extraction factors only; these are the useful indicators. If a particular indicator has a low communality (less than 0.5) or a larger communality (more than 0.9), then one should consider dropping it from the analysis (Brace et al. 2006).

Communalities are presented in Table 4.3. No communalities for all indicators in both groups had a value lower than 0.5 or larger than 0.9. This implies that PCA explains much of the associated variance for all items. In other words, indicators have much in common with each other and are very closely related. Communalities for LSELU are lower than for HSELU, while communalities for All-HEI show the lowest values among four samples and communalities for Outsiders are the largest among four samples.

Table 4.3 Communalities of four samples

Latent variable/item	Initial	LSELU extraction	HSELU extraction	All-HEI extraction	Outsiders extraction
I. Universal values					
1Worldatpeace	1	0.684	0.721	0.576	0.711
2Influential	1	0.723	0.761	0.638	0.756
3Ambitious	1	0.741	0.771	0.693	0.782
4Broadminded	1	0.652	0.680	0.616	0.620
5Authority	1	0.698	0.692	0.673	0.711
6Creativity	1	0.695	0.745	0.520	0.679
7Socialpower	1	0.692	0.745	0.634	0.737
8Socialorder	1	0.651	0.851	0.545	0.693
9Prevention	1	0.780	0.759	0.695	0.784
10Variedlife	1	0.700	0.769	0.697	0.764
11Socialjustice	1	0.690	0.740	0.623	0.782
12Enjoyinglife	1	0.650	0.831	0.664	0.802
13Selfdiscipline	1	0.708	0.805	0.526	0.784
14Unitywnature	1	0.555	0.737	0.705	0.789
15Health	1	0.831	0.813	0.775	0.799
16Responsible	1	0.810	0.792	0.751	0.814
17Respectful	1	0.756	0.824	0.713	0.868
18Moderate	1	0.655	0.730	0.658	0.743
19Equality	1	0.687	0.733	0.660	0.713
20Acceptinglife	1	0.691	0.793	0.635	0.810
21Choosingoals	1	0.695	0.683	0.684	0.741
II. Awareness of consequences					
22aClimateyou	1	0.803	0.838	0.767	0.819
23bClimatecountry	1	0.753	0.785	0.716	0.743
24cClimateplants	1	0.772	0.831	0.668	0.815
25aForestyou	1	0.626	0.843	0.737	0.774
26bForestcountry	1	0.729	0.720	0.717	0.781
27cForestplants	1	0.748	0.876	0.705	0.784
28aToxicyou	1	0.634	0.856	0.748	0.860
29bToxicountry	1	0.700	0.854	0.752	0.812
30cToxicplants	1	0.715	0.855	0.791	0.857
III. Ascription of responsibility					
31Government1	1	0.703	0.701	0.669	0.839
32Ifeelobligation	1	0.731	0.764	0.684	0.743
33Ifeelasense	1	0.684	0.739	0.739	0.777
34Business	1	0.701	0.800	0.683	0.851
35Government2	1	0.771	0.780	0.691	0.844
36Government3	1	0.800	0.826	0.767	0.782

(continued)

Table 4.3 (continued)

Latent variable/item	Initial	LSELU extraction	HSELU extraction	All-HEI extraction	Outsiders extraction
37Companies	1	0.754	0.728	0.617	0.823
38Peoplelike	1	0.711	0.744	0.625	0.793
39Industry	1	0.717	0.700	0.515	0.767
IV. Personal intelligences					
40Anticipatesobsta	1	0.761	0.635	0.639	0.715
41Adaptsideas	1	0.734	0.750	0.632	0.755
42Solicitsinput	1	0.647	0.757	0.624	0.661
43Takesrisks	1	0.674	0.682	0.561	0.808
44Relateswell	1	0.727	0.668	0.648	0.738
45Stayscomposed	1	0.731	0.740	0.627	0.722
46Leadsbyexam	1	0.737	0.725	0.609	0.782
47Advocateschange	1	0.640	0.742	0.619	0.712
48Actsimpulsively	1	0.768	0.764	0.746	0.758
49Personallyleads	1	0.761	0.823	0.677	0.784
50Keepspromi	1	0.668	0.710	0.662	0.769
51Acknowledgesmist	1	0.681	0.680	0.644	0.833
52Articulatesacompe	1	0.653	0.714	0.623	0.832
53Canseethings	1	0.736	0.792	0.694	0.759
54Believescapable	1	0.758	0.719	0.675	0.796
55Bendsrules	1	0.591	0.751	0.592	0.688
56Doubtsownability	1	0.718	0.776	0.701	0.772
57Establishesclose	1	0.660	0.706	0.598	0.673
58Hesitatestoact	1	0.771	0.740	0.731	0.750
59Changesstrategy	1	0.650	0.802	0.676	0.788
IV. Demographics					
63Gender	1	0.755	0.690	0.623	0.708
64Age	1	0.752	0.760	0.656	0.739

Extraction method: principal component analysis

4.2.1.5 Eigenvalues and Explained Variance of Four Samples

These explain a percentage of all the variance, and the cumulative percentage. Components are ranked in order of how much variance each accounts for. This is the first part of the output that gives a clear indication of the solution, in terms of how many factors explain how much variance. Previous tables and matrices are important, though, in indicating whether the solution is likely to be a good one.

Table 4.4 presents eigenvalues for the entire analysis in a concise manner, as well as estimations of explained variance for a final solution of the PCA calculation. This table contains two sets of results. The section entitled "Initial Eigenvalues" presents own values, percentage of variance, and cumulative percentage of variance

Table 4.4 Total variance explained in four samples

Samples	Components	Initial eigenvalues			Extraction sums of squared loadings		
		Total	% of variance	Cumulative %	Total	% of variance	Cumulative %
LSELU	1	8.942	14.659	14.659	8.942	14.659	14.659
	18	1.022	1.676	71.098	1.022	1.676	71.098
	61	0.044	0.072	100.000			
HSELU	1	8.390	13.754	13.754	8.390	13.754	13.754
	20	1.006	1.649	75.971	1.006	1.649	75.971
	61	0.035	0.058	100.000			
All-HEI	1	10.167	16.668	16.668	10.167	16.668	16.668
	17	1.037	1.700	66.443	1.037	1.700	66.443
	56	0.194	0.318	98.809			
Outsiders	1	10.034	16.450	16.450	10.034	16.450	16.450
	17	1.004	1.646	76.862	1.004	1.646	76.862
	61	0.014	0.022	100.000			

for each factor ranked in the magnitude of eigenvalues. In LSELU sample, the first eigenvalue is 8.94, and this explains 14.7 % of variance. Eigenvalues are greater than zero and their sum is 61. The section entitled "Extraction Sums of Squared Loadings" reproduces the number of extracted factors in the PCA (18 for LSELU sample). Sums of squared saturations are identical to eigenvalues, and 18 factors explain 71 % of variance. In HSELU sample, the number of extracted factors is 20 and explains 76 % of variance. In All-HEI sample, the number of extracted factors is 17 and explains 66 % of variance. Finally, in Outsiders sample, the number of extracted factors is 17 and explains 77 % of variance.

4.2.1.6 Scree Plot of Four Samples

This is an alternative to eigenvalues >1.0, to decide which component should be extracted. The eigenvalues are plotted in decreasing order. This is called a scree plot because the shape of the curve is reminiscent of the profile of scree which accumulates at the foot of steep hills (Gardner 2003; Brace et al. 2006). We are trying to distinguish the "mountain" (i.e., principal components based on true covariance) from "rocks" (i.e., principal components based on random error) (Gardner 2003). No figure is presented in this section for scree plot. This graph allows for determining the number of factors that best represent any significant variance described by the correlation matrix. However, inspection of graphs suggests that 18, 20, 17, and 17 factors explain the main significant variance of the correlation matrix according LSELU, HSELU, All-HEI, and Outsiders samples, respectively. This decision is based on the position of the "elbow" in graphs. In LSELU sample, it is at factor eighteen, suggesting that the amount of variance explained by 18 factors and subsequent factors is low and virtually equivalent to that determined by the eigenvalue method. The same explanation is given for the other three samples according to above-mentioned factors.

4.2.1.7 Component Matrix of Four Samples

The matrix of initial factors is the matrix for principal component factors. It is a structural matrix because it involves correlations of each indicator with each principal component. This is a table of the factor loadings. Each column shows the loading of each indicator for that component. Loading can be thought of as the correlation between the component and the indicator: thus, the larger the number, the more likely it is that the component underlies that variable. The extent of communality indicates how much of that indicator's variance is explained by the solution to the factor analysis.

The decision concerning the number of factors in the component matrix is based on the eigenvalue rule of 1, not on results of previous scree plots, although in this study both methods give the same results as shown in Table 4.4.

An inspection of the component matrix reveals that factors have positive and negative values across all variables. In other words, factors are a combination of positive and negative saturation in the component matrix. In the corresponding table, loadings are ranked by component. The first four components for four samples will be shown in Table 4.5. These loadings may be useful for seeing the pattern of which indicator loads most strongly with which factors. In particular, the negative loadings found here may be an artifact of the calculation method. Blanks are very low loadings.

4.2.1.8 Matrix of Residual Correlation for Four Samples

This represents the difference of each value between the matrix of reproduced correlation (communalities) and the matrix of observed correlation. In this study, the residual correlation matrix is not shown because of the extension. The small size of most of the residuals is another indication of factorability and is also an indication of a good factor analysis solution (Brace et al. 2006).

Usually when the matrix of residual correlation is generated, the matrix of reproduced correlations is obtained. However, this matrix is identical to communalities presented in Table 4.3, and therefore, they will not be presented in this section. The matrix of reproduced correlations was calculated by the equation from the fundamental theorem (see Appendix D.1) submitted to factorial saturation in the correlation matrix.

In determining how well PCA explains the observed matrix of correlations, the matrix of residual correlation was calculated by subtracting each value from reproduced correlations to the corresponding value in the matrix of correlations. This produces the matrix of residual correlation. These residual values are close to

Table 4.5 Pattern of the first four components found by the PCA and the representative relations of the latent variables

	Components for LSELU sample				Components for HSELU sample				Components for All-HEI sample				Components for Outsiders sample			
	Items or indicators				Items or indicators				Items or indicators				Items or indicators			
PCA values	1	2	3	4	1	2	3	4	1	2	3	4	1	2	3	4
Higher (0.7–0.8)															45^{IV}	
Middle (0.6–0.69)	6^{I}	22^{II} 24^{II} 26^{II}			9^{I} 14^{I} 16^{I} 17^{II} 25^{II} 32^{III}				25^{II}				16^{I} 17^{I} 19^{I} 30^{II}		46^{IV} 51^{IV} 53^{IV}	
Lower (0.4–0.59)	4^{I} 8^{I} 9^{I} 13^{I} 14^{I} 15^{I} 16^{I} 17^{I} 18^{I} 19^{I} 21^{I} 32^{III} 33^{III} 38^{III}	23^{II} 25^{II} 28^{II} 29^{II} 30^{II} 32^{II} 34^{II} 35^{II} 36^{II} 38^{II}	1^{I} 15^{I} 17^{I} 19^{I} 41^{IV} 42^{IV} 44^{IV} 52^{IV}	5^{I} 7^{I} 48^{IV} 55^{IV} 56^{IV} 58^{IV}	1^{I} 4^{I} 11^{I} 15^{I} 19^{I} 21^{I} 22^{II} 23^{II} 26^{II} 28^{II} 29^{II} 31^{III} 33^{III} 35^{III} 36^{III}	1^{I} 4^{I} 19^{I} 46^{IV} 47^{IV} 54^{IV}	24^{II} 30^{II} 43^{IV} 44^{IV} 45^{IV} 52^{IV} 59^{IV}	48^{IV} 56^{IV}	4^{I} 6^{I} 8^{I} 9^{I} 13^{I} 14^{I} 15^{I} 16^{I} 17^{I} 18^{I} 21^{I} 22^{II} 23^{II} 26^{II} 28^{II} 29^{II}	24^{II} 27^{II} 30^{II} 34^{III} 36^{III} 45^{IV} 52^{IV} 54^{IV}	12^{I}	5^{I} 55^{IV} 59^{IV}	1^{I} 4^{I} 6^{I} 9^{I} 11^{I} 13^{I} 14^{I} 15^{I} 18^{I} 22^{II} 23^{II} 24^{II} 25^{II} 26^{II} 27^{II}	47^{I} 34^{III} 35^{III} 37^{III} 40^{IV} 41^{IV} 44^{IV} 47^{IV} 50^{IV} 52^{IV} 54^{IV} 57^{IV}	9^{I} 14^{I} 15^{I} 16^{I} 17^{I} 18^{I} 42^{IV} 52^{IV} 58^{IV}	55^{IV}

(continued)

Table 4.5 (continued)

	Components for LSELU sample				Components for HSELU sample				Components for All-HEI sample				Components for Outsiders sample			
	1	2	3	4	1	2	3	4	1	2	3	4	1	2	3	4
					38^{III}				32^{III}				28^{II}			
									33^{III}				29^{II}			
									35^{III}				31^{III}			
									38^{III}				32^{III}			
									56^{IV}				33^{III}			
													34^{III}			
													35^{III}			
													36^{III}			
													37^{III}			
													38^{III}			
													39^{III}			

Note Latent variables are: [I] universal values, [II] awareness of consequences, [III] ascription of responsibility, [IV] personal intelligences. Numbers are indicators or items; indicator names are in Table 4.1. PCA values = saturation values extracted by principal component analysis method. *Indicators* 10, 20, 49 were not selected for anyone zero. To give a rough idea of how to fit the adjustment, statistical software counts the number of non-redundant residuals whose absolute value is greater than 0.05.

4.2.1.9 Matrix of Rotated Factors of Four Samples

This represents the matrix of initial factors which has been rotated to produce a solution which is easier to interpret. It was not possible to obtain a rotation converged in 25 iterations for components and indicators for four samples.

Twenty-five iterations were carried out with a varimax rotation using *SPSS20* statistical software (Arbuckle 2012), but this was insufficient to obtain loading values for four samples. The purpose of the rotation is to produce an easier solution for interpreting data. The rationale of rotation criteria is based on continuing the rotation until the squared sum of factorial saturation variances for each factor is as large as possible (Gardner 2003). Failure to obtain a rotated matrix does not alter initial results; the only difference is the frame of reference used to describe the location of points in space.

4.2.2 Estimation of Factor Scores

This section describes the pattern of components that shows the representative relation of latent variables which were the leading factors underlying behavior for sustainability in LSELU, HSELU, All-HEI, and Outsiders samples.

4.2.3 Pattern of Components and Latent Variables for Four Samples

Data are analyzed by means of a principal component analysis, and outcomes of the underlying latent variables are presented in order to interpret PCA data, the following authors were used: Jollife (1986), Basilevsky (1994), Gardner (2003), and Brace et al. (2006).

Table 4.5 shows the pattern found as a representative relation of latent variables, which were the leading factors underlying behavior for sustainability in LSELU, HSELU, All-HEI, and Outsiders samples. In the left-most column, saturation values from the component matrix extracted using the PCA method are presented. On the right side, four columns are grouped for each sample: LSELU, HSELU, All-HEI, and Outsiders. Each sample consists of 4 columns composed of the first four components of the initial matrix. For high saturation value (0.70–0.80) at the Outsiders sample, indicator 45Staycomposed, pertaining to the latent variable "personal intelligences," appears. There are no data for this high saturation value for LSELU, HSELU, and All-HEI samples.

For middle saturation values (0.60–0.69) related to the first component, the LSELU sample loads indicator 6Creativity, pertaining to the latent variable "universal values." For the HSELU sample loads six indicators, four (9Prevention, 14Unitywnature, 16Responsible, 17Respectful) pertaining to latent variables "universal values," one (25aForestyou) pertaining to "awareness of consequences," and one (32Feelobligation) pertaining to "ascription of responsibility." For the All-HEI sample loads only item 25aForestyou, pertaining to the latent variable "awareness of consequences." For the Outsiders sample loads four indicators three (16Responsible, 17Respectful, and 19Equality) pertaining to latent variables "universal values" and one indicator (30cToxicplants) pertains to the latent variable "awareness of consequences."

The lower saturation values (0.40–0.59) for the first component at LSELU sample load 11 indicators (4Broadminded, 8Socialpower, 9Prevention, 13Selfdiscipline, 14Unitywnature, 15Health, 16Responsible, 17Respectful, 18Moderate, 19Equality, 21Choosingoals) associated to "universal values" and three indicators (32Ifeelobligation, 33Ifeelsense, 38Peoplelike) associated to "ascription of responsibility." For the HSELU sample loads six indicators (1Worldatpeace, 4Broadminded, 11Socialjustice, 15Health, 19Equality, 21Choosingoals) associated to "universal values," five (22aClimateyou, 23bClimatecountry, 26bForestcountry, 28aToxicyou, 29bToxicountry) associated to "awareness of consequences," and five (31Government1, 33Ifeelsense, 35Government2, 36Government3, 38Peoplelike) associated to "ascription of responsibility." For the All-HEI sample loads eleven indicators (4Broadminded, 6 Creativity, 8Socialpower, 9Prevention, 13Selfdiscipline, 14Unitywnature, 15Health, 16Responsible, 17Respectful, 18Moderate, 2121Choosingoals) associated to "universal values," five (22aClimateyou, 23bClimatecountry, 26bForestcountry, 28atoxicyou, 29bToxicountry) to "awareness of consequences," four (32Ifeelobligation, 33Ifeelsense, 35Government2, 38Peoplelike) to "ascription of responsibility," and one (56Doubstownability) to "personal intelligences." For the Outsiders sample loads nine indicators (1Worldatpeace, 4Broadminded, 6 Creativity, 9Prevention, 11Socialjustice, 13Selfdiscipline, 14Unitywnature, 15Health, 18Moderate) associated to "universal values," eight out of nine indicators (22aClimateyou, 23bClimatecountry, 24cClimateplants, 25aForestyou, 26bForestcountry, 27cForestplants, 28atoxicyou, 29bToxicountry) associated to "awareness of consequences," and the entire list of indicators associated to "ascription of responsibility."

"Universal values" appear more frequently than the latent variable "awareness of consequences" for middle saturation values related to the first component in all samples, while "ascription of responsibility" and "personal intelligences" appear more frequently in lower saturation values for second and third component than the other two latent variables "universal values" and "awareness of consequences." "Demographics" did not appear in any saturation values for any sample in the first four components.

4.3 Results of a Confirmatory Second-Order Factor Analysis

CFA results appear according to the following issues: (1) identification of the model which entails a path diagram that show the latent structure of the SB second-order model to be characterized by four first-order factors, and (2) estimation of CFA model parameters.

4.3.1 Identification of the CFA Model

A CFA (for more details on calculations, see Appendix D.2) was conducted with *Mplus* (version 6.0; for further discussion see Muthén and Muthén 1998–2010). CFA was used to guide the specification and evaluation of the factor model based on a strong empirical, conceptual foundation to validate such SB construct (Asparouhov and Muthén 2014; Brown 2006). Based on the prior section of this chapter (and in a preliminary analysis: Juárez-Nájera 2010) and theory bearing on the Stern et al. (1999) model of value-belief-norm, and on Schwartz (1994), Hines et al. (1987/88), Gardner (2001) and Corral-Verdugo (2010) frameworks (see Fig. 3.1), a second-order model was specified.

A schematic representation of the second-order model is presented in Fig. 4.1 to illustrate the concepts of parameter estimation. Figure 4.1 notation was selected in this study following the conventions of factor analysis and structural equation modeling. Second-order factor is symbolized by Ksi1 ($\xi1$) with 1 subscript as one single factor. Four first-factor latent variables are symbolized by Eta (η) with 4 subscripts ($\eta1$ to $\eta4$). Factor variances and covariances are notated by phi (ϕ) with 4 subscripts ($\phi1$ to $\phi4$) to each latent factor. The unidirectional arrows from the factors (e.g., $\xi1$, $\eta1$) to the indicators (e.g., $X1$, $Y1$) depict direct effects (regressions) of the latent dimensions onto the observed measures (X are exogenous variables, and Y are endogenous, dependent variables); the specific regression coefficients or factor loadings are the lambdas (λ). The error variances are symbolized by epsilon (ε) in each factor.

The factor loadings (lambdas-λ) in Fig. 4.1 also possess numerical subscripts to point out the specific indicators ($X1$ or $Y1$) of the relevant matrices. For example, λ_Y11 indicates that the $Y1$ measure loads on the first endogenous factor ($\eta1$), and λ_X11 indicates that the $X1$ measure loads on the first exogenous factor ($\xi1$). This numeric notation assumes that the indicators were ordered ($Y1 \ldots Y59$ and $X1, X2, X3$) in the input variance–covariance matrix. The input data are depicted in the path diagram 4.2.

With these notions in mind, the path diagram of Fig. 4.2 shows the latent structure of the SB second-order model to be characterized by four first-order factors in which $Y1$ to $Y21$ loaded onto the latent variable of universal values, $Y22$ to $Y30$ loaded onto the latent variable of awareness of consequences, $Y22$ to $Y30$

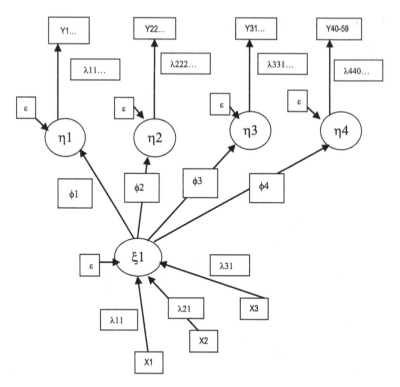

Fig. 4.1 Schematic representation of a second-order (Ksi-$\xi1$), four-factor (Eta-$\eta1$ to Eta-$\eta4$) CFA model. 59 latent Y endogenous variables, three X exogenous variables, factor loadings (lambda λ), factor variances (Phi $\phi1$ to $\phi4$) and error variances (ε) are depicted

loaded onto the latent variable of ascription of responsibility, and $Y40$ to $Y59$ loaded onto the latent variable of personal intelligences. The four factors are presumed to be intercorrelated (prior section and Juárez-Nájera et al. 2010). The empirical feasibility of the second-order model should be evidenced by the patterning of correlations among factors in the first-order model. $Y1$, $Y22$, $Y31$, and $Y40$ were used as marker indicators for universal values, awareness of consequences, ascription of responsibility, and personal intelligences, respectively.

4.3.2 Estimation of the CFA Model Parameters

As shown in Fig. 4.2, a basic path model was tested using 59 indicators of sustainable behavior and an $N = 327$ adults in four higher educational institutions.

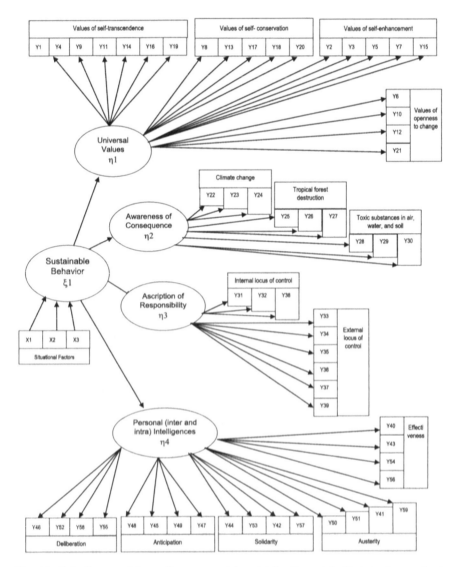

Fig. 4.2 Path diagram of the confirmatory model containing the sustainable behavior construct. Y1 to Y21 denote the universal values (η1), Y22 to Y30 denote awareness of consequences (η2), Y31 to Y39 denote ascription of responsibility (η3), and Y40 to Y59 denote personal intelligences (η4). The complete list of X and Y indicators is shown in Table 4.1

Table 4.6 shows the results for LSELU, HSELU, All-HEI, and Outsiders samples. That is, results for a sustainable behavior second-order factor model among 127 participants from a university in a country (Mexico) with a lower socioeconomic level (LSELU), 105 participants from universities in three countries (Germany, Switzerland, Canada) with higher socioeconomic level (HSELU), 232 participants

Table 4.6 Results of chi-square (χ^2) and other goodness-of-fit indices for a sustainable behavior second-order factor model of different samples

	χ^2	df	χ^2 diff	RMSEA (90 % CI)	CFIT	WRMSR	TLI	CFI
LSELU $N = 127$	2262.921	1947	4652.532	0.040 (0.033–0.046)	0.997	1.205	0.853	0.858
HSELU $N = 105$	2473.628	1996	3599.038	0.048 (0.041–0.054)	0.721	1.303	0.678	0.689
All-HEI $N = 232$	2772.262	1996	6935.441	0.041(0.037–0.045)	1.000	1.345	0.835	0.841
Outsiders* $N = 95$	2578.717	1938	10263.273	0.059 (0.053–0.065)	0.01	1.412	0.920	0.922

LSELU university in a country with a lower socioeconomic level, *HSELU* 3 universities in 3 countries with higher socioeconomic level, *All-HEI* members of all universities among countries (lower socioeconomic level and higher socioeconomic level), Outsiders of higher education institutions

df degrees of freedom, *diff* from the baseline model, *RMSEA* root-mean-square error of approximation, *CI* confidence interval, *CFIT* test of close fit, *WRMSR* weighted root-mean-square residual, *TLI* Tucker–Lewis index, *CFI* comparative fit index

*The residual covariance matrix and the latent variable covariance matrix are not positive definite problem involving variable Y24 (climate change) and variable eta2 (awareness of consequence)

from the four universities mentioned in Chap. 3, and 95 participants who are outsiders of the Mexican university (see section of data collection procedure).

Guided by suggestions provided in Brown (2006: 87), a description of goodness-of-fit indices in Table 4.6 is as follows:

- Chi-square (χ^2) evaluates the reasonability of the hypothesis that $S = \Sigma$ (S is the input matrix and Σ is the predicted covariance or correlation matrix; see Appendix D for more details); the critical χ^2 value ($\alpha = 0.05$) is $1.96^2 = 3.8416$. The model for each sample exceeds this critical value and the null hypothesis that $S = \Sigma$ is rejected. Thus, a statistically significant variance supports the alternate hypothesis, meaning that the model does not fit the data well or the model estimates do not sufficiently reproduce the sample variances and covariances.

- The df is the degrees of freedom or the difference in the number of knowns (number of variances and covariances in the input matrix) and the number of unknowns (the number of freely estimated model parameters). In this case, the model for each sample depicted an overidentified model or positive values greater than 1. Accordingly, the model was more than 1900 degrees of freedom (df) from *Mplus* results and supported by 1830 df from KMO and Bartlett's test in PCA Table 4.2, except for Outsiders sample with less df (1711).

- χ^2 test of model fit for the baseline model is χ^2 diff.

- Root-mean-square error of approximation (RMSEA) and its 90 % confidence interval (90 % CI) assess the extent to which a model fits reasonably well in the population. That is, an acceptable model fit was defined by allowing the correlations among the factors to be freely estimated. As per the rule of thumb, RMSEA values less than 0.08 suggest adequate model fit (a "reasonable error of approximation"), RMSEA values less than 0.05 suggest good model fit, and that model with RMSEA greater or equal than 0.1 should be rejected. Outputs for all samples suggest good model fit.

- Test of close fit (CFIT) is the probability RMSEA ≤0.05.
- Maximum likelihood (ML) allows of how closely do the correlations among the indicators predicted by the factor analysis parameters approximate the relationships seen in the input correlation matrix of multivariate normal distribution of the variables, that is, goodness of fit. Goodness of fit in this research was evaluated using the weighted root-mean-square residual (WRMSR). Brown (2006: 76) highlights that if one or more of the factor indicators are categorical (as it was the case in this study; e.g., five Likert values in universal value section from 1: totally agree to 5: totally disagree), normal theory of ML should not be used because of underestimate standard errors, thus increasing the risk of type I error.
- The Tucker–Lewis index (TLI) is determining the suitability of the factor model and the number of latent variables. That is, TLI has features that compensates for the effect of model complexity. TLI is non-normed, which means that its values can fall outside the range of 0.0–1.0. TLI values in the range of 0.90–0.95 may be indicative of acceptable model fit. However, when fit indices fall in these "marginal" ranges (as LSELU, HSELU, and All-HEI samples), it is especially important to consider the consistency of model fit as expressed by the various types of fit indices in tandem with the particular aspects of the analytic situation, for example, when N is somewhat small (less than 100).
- Comparative fit index (CFI) evaluates the fit of a user-specified solution in relation to a more restricted, nested baseline model. The CFI has a range of possible values of 0.0–1.0, with values closer to 1.0 implying good model fit. LSELU, All-HEI, and Outsiders samples have good model fit. HSELU has moderately good model fit.

On the basis of the results of the overall goodness-of-fit indices, it can be concluded that the four-factor model evidence fits the data.

Inspection of the standardized residuals and modification indices indicated no localized points of ill fit in the solution. Unstandardized and completely standardized parameter estimates from this solution are not shown in this section, but the table is available upon request to the author. All freely estimated unstandardized parameters were statistically significant (ps ≤ 0.05). R-square revealed that the indicators were moderately related to their purported latent factors as Figs. 4.3, 4.4, 4.5, and 4.6 explain, respectively, LSELU, HSELU, All-HEI, and Outsiders samples.

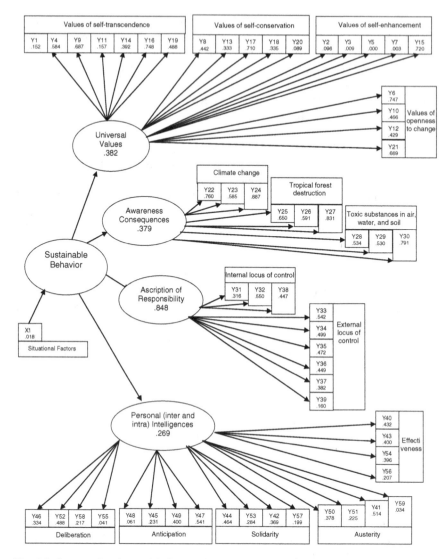

Fig. 4.3 *R*-square for observed indicators, below their identification number, of the confirmatory model for LSELU sample. The complete list of *X* and *Y* indicators is shown in Table 4.1 and Appendix C

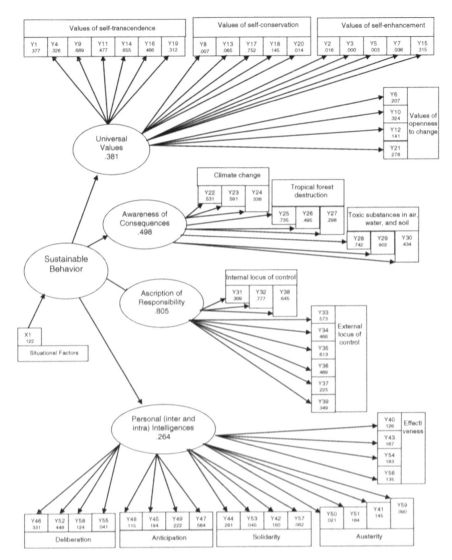

Fig. 4.4 R-square for observed indicators, below their identification number, of the confirmatory model for HSELU sample. The complete list of X and Y indicators is shown in Table 4.1 and Appendix C

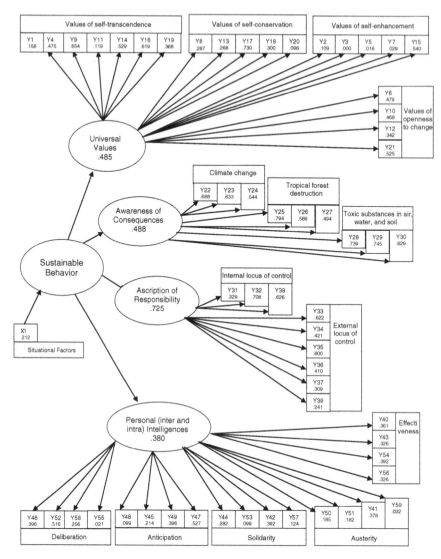

Fig. 4.5 *R*-square for observed indicators, below their identification number, of the confirmatory model for All-HEI sample. The complete list of *X* and *Y* indicators is shown in Table 4.1 and Appendix C

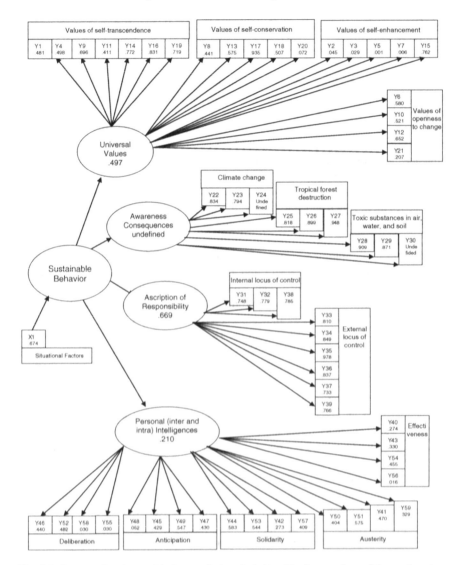

Fig. 4.6 *R*-square for observed indicators, below their identification number, of the confirmatory model for Outsiders sample. The complete list of *X* and *Y* indicators is shown in Table 4.1 and Appendix C

4.4 Summary of Findings in Sustainable Behavior Pathways

In the following section, the analysis focuses on the behavioral domains that determine universal values, awareness of consequences, ascription of responsibility, and personal intelligences as latent variables of sustainable behavior. The goal is to identify which latent variable explains the willingness to sustainable behavior, and more specifically, which perceptions that are currently most relevant and likely to influence behavioral change in members of higher educational institutions compared to outsiders of HEI.

The data were analyzed first by means of a principal component analysis with a varimax rotation which did not converge after 25 iterations at LSELU, HSELU, All-HEI, and Outsiders samples. Cronbach's alpha for any sample was not calculated because Brown (2006: 239) states that it is a mis-estimator of the scale reliability of a multiple-item questionnaire. Three indicators of factorability load a good factorability level: the KMO and Bartlett tests show that the data have a probability of factorability; also high values in communalities, and the very low residual values from the matrix of reproduced correlations indicate optimum outcomes. The rest of factorability indicators loads a moderately factorability level.

Within the data, for LSELU sample 18, components were found with an eigenvalue of greater than 1.0; they accounted 71 % of the explained variance. Scree plots indicate 18 components, very close to the eigenvalue. For HSELU sample, 20 components were found with an eigenvalue of greater than 1.0; they accounted 76 % of the explained variance. For All-HEI sample, 17 components were found with an eigenvalue of greater than 1.0; they accounted 66 % of the explained variance. Scree plots indicate 17 components, very close to the eigenvalue. Finally, for Outsiders sample, 17 components were found with an eigenvalue of greater than 1.0; they accounted 77 % of the explained variance. Scree plots indicate 17 components, very close to the eigenvalue.

The communalities values in all samples were an average of 0.7; this means that the variables had much in common with each other with 127, 105, 232, and 95 subjects participated. That is, the associated variance from 66 to 77 % as opposed to 40–60 % from previously mentioned models; therefore, this is a very promising model depicted by the PCA statistical procedure.

Regarding the latent variables with saturation values related to the first four components, "universal values" appear more frequently than "awareness of consequences" for middle saturation values related to the first component in all samples, while "ascription of responsibility" and "personal intelligences" appear more frequently in lower saturation values for second and third component than "universal values" and "awareness of consequences." "Demographics" did not appear in any saturation values for any sample in the first four components.

On the CFA, Figs. 4.3, 4.4, 4.5, and 4.6 showed, respectively, the expectations of HEI's members and outsiders within each latent variables and their degree of influence. The regression analysis showed an index of determination of R-square for each latent variable and their indicators.

The results point out that "ascription of responsibility" (refers to people's inclination to accept for the consequences of their behavioral choices toward the welfare of others) accounted for 84.8 % of the explained variance among LSELU members ($N = 127$), and 80.5 % for HSELU members ($N = 105$). For all university members, the R-square over SB was less high, accounted 72.5 % of the explained variance with $N = 232$, and even lower for outsiders with $N = 95$, accounted 66.9 % of the explained variance. Within this latent variable, the beliefs were classified into two groups according to the arrangements of internal locus of control (represents an individual's perception of whether he/she has the skills to provoke changes through his/her own behavior) and external locus of control (refers to concepts based on the belief of some individuals do not intend to provoke change because they attribute change to randomness or other powerful forces), as presented in Table 3.2. This grouping resulted in an index of determination of R-square that accounted for outsiders from 75 to 98 % of the explained variance either the internal locus of control or the external locus of control. The internal locus of control accounted among LSELU members from 31 to 55 % of the explained variance and the external locus of control from 16 to 54 %. The internal locus of control accounted among HSELU members from 31 to 78 % of the explained variance and the external locus of control from 22 to 61 %. Finally, the internal locus of control accounted among all university members from 33 to 71 % of the explained variance and the external locus of control from 24 to 62 %.

The results indicate that "universal values" (represent conscious goals as response to needs of individuals, coordinated interaction and smooth functioning and survival of groups) accounted for 38.2 % of the explained variance among LSELU members, and 38.1 % for HSELU members. All university members accounted 48.5 % of the explained variance, and outsiders accounted 49.7 % of the explained variance. Within this latent variable, the beliefs were classified in four groups: self-transcendence, conservation, self-enhancement, and openness to change, as depicted in Fig. 2.4. This grouping resulted in an index of determination of R-square that accounted predominantly in all samples for values of self-transcendence (universalism and benevolence; see Table 2.2) and to a lesser extent values of self-enhancement (hedonism, achievement, and power; see Table 2.2). According to value theory, the structure of dynamic relationships among four groups shows that member actions which express values of self-transcendence were likely to be in conflict with those which express values of self-enhancement.

The results indicate that "awareness of consequences" (refers to a person's receptivity for cues signaling situational needs) accounted for 37.9 % of the explained variance among LSELU members, and 49.81 % for HSELU members. All university members accounted 48.8 % of the explained variance, and outsiders accounted undefined of the explained variance. Within this latent variable, the beliefs were classified in three groups: climate change, tropical forest destruction, and toxic substances in the air, water, and soil referred to a problem for one-self and one's family, for the whole country, and for other plant and animal species, as presented in Table 3.2. This grouping resulted in an index of determination of R-square that accounted for LSELU members from 79–89 % of the explained variance when the problem is for other species and in a lesser extent from 53 to 58 % of the explained variance when the problem is for the country. HSELU members accounted 53–74 % of the explained variance when the problem is for one-self and in a lesser extent from 34 to 43 % when the problem is for other species; all university members accounted from 54 to 63 % of the explained variance when the problem is for other species and in a lesser extent from 69 to 79 % of the explained variance when the problem is for one-self, and finally, for the outsiders accounted high values up to 79 % of the explained variance but two indicators were undefined.

The results show that "inter- and intrapersonal intelligences" (psychological features concerned with the capacity to understand the intentions, motivations, and desires of others and one-self associated to sustainability) accounted for 26.9 % of the explained variance among LSELU members, and 26.4 % for HSELU members. All university members accounted 38 % of the explained variance and outsiders accounted 21 % of the explained variance. Within this latent variable, the actions were classified into five groups: effectiveness (the tendency to respond swiftly to demands), anticipation (the expectation of future actions), austerity (the conservative behavior in the face of an uncertain world), deliberation (the act of directing actions toward a specific end), and solidarity (the tendency to be concerned about and to act in favor of others), as presented in Table 3.3. This grouping resulted in an index of determination of R-square that accounted for LSELU members from 20 to 43 % of the explained variance for effectiveness, from 6 to 54 % of the explained variance for anticipation, from 20 to 46 % of the explained variance for solidarity, from 3 to 51 % of the explained variance for austerity, and 4–49 % of the explained variance for deliberation. HSELU members accounted from 12 to 18 % of the explained variance for effectiveness, from 11 to 56 % of the explained variance for anticipation, from 4 to 26 % of the explained variance for solidarity, from 2 to 18 % of the explained variance for austerity, and 4–45 % of the explained variance for deliberation. All university members accounted from 32 to 39 % of the explained variance for effectiveness, from 10 to 53 % of the explained variance for anticipation, from 10 to 39 % of the explained variance for solidarity, from 3 to 18 % of the explained variance for austerity, and from 2 to 51 % of the explained variance for deliberation. Finally, outsiders accounted from 2 to 46 % of the explained variance for effectiveness, from 5 to 55 % of the explained variance for anticipation, from 27 to 58 % of the explained variance for solidarity, from 33 to 58 % of the explained variance for austerity, and 3–48 % of the explained variance for deliberation.

Estimates from the second-order four-factor solution indicate moderate relationships of the latent variable "ascription of responsibility," while "universal values," "awareness of consequences," and "personal intelligences" were poorly related in accord with previous evidence and theory, less than 30 %. However, these results are consistent with the position that the Stern et al. model scales are reliable indicators of the construct of sustainable behavior because of the use of empirical data and strong procedural statistics, PCA and CFA.

On a policy basis, in order to encourage sustainability in higher education "universal values," "awareness of consequences" and "personal skills" must be fostered, while "ascription of responsibility" is presented as principal determinant among all kind of members at an HEI regardless of the socioeconomic structure of the nation in which the HEI is located, even other type of people who is outsider to any university education must be considered. Future research should consider additional HEI and a greater number of participants.

References

Arbuckle, J. L. (2012). *User's Guide: IBM-SPSS Amos 20*. Amos Development Corporation.

Asparouhov, T., & Muthén, B. (2014). Multiple-group factor analysis alignment. *Structural Equation Modeling: A Multidisciplinary Journal, 21*, 1–14. doi:10.1080/10705511.2014. 919210.

Barendse, M. T., Oort, F. J., & Timmerman, M. E. (2014). Using exploratory factor analysis to determine the dimensionality of discrete responses. *Structural Equation Modeling: A Multidisciplinary Journal*. doi:10.1080/10705511.2014.934850.

Bartholomew, D. J. (1987). *Latent variable models and factor analysis*. London: Charles Griffin & Company LTD/New York: Oxford University Press.

Basilevsky, A. (1994). Statistical factor analysis and related methods. Theory and applications. In W. A. Shewhart & S. S. Wilks (Eds.), *Probability and mathematical statistics*. New York: Wiley.

Bollen, K. A. (2002). Latent variables in psychology and the social sciences. *Annual Review of Psychology, 53*, 605–634.

Bollen, K. A. (2007). An overview of structural equation models with latent variables. In *Proceedings of the Symposium on Computational Research at Miami University*, Miami, USA, March 1–2, 2007.

Bollen, K. A., & Lenox, R. (1991). Conventional wisdom on measurement: A structural equation perspective. *Psychological Bulletin, 110*(2), 305–314.

Brace, N., Kemp, R., & Snelgar, R. (2006). *SPSS for psychologist: A guide to data analysis using SPSS for Windows version 12 & 13*. New York: Palgrave Macmillan.

Brown, T. A. (2006). *Confirmatory factor analysis for applied research*. NY: The Guilford Press.

Corral-Verdugo, V. (2010). *Psicología de la sustentabilidad: un análisis que nos hace pro ecológicos y pro sociales*. México: Trillas.

Dunn, E. C., Masyn, K. E., Jones, S. M., Subramanian, S. V., & Koenen, K. C. (2014). Measuring psychosocial environments using individual responses: An application of multilevel factor analysis to examining students in schools. *Prevention Science*. doi:10.1007/s11121-014-0523-x.

Gardner, H. (2001). *Estructuras de la Mente. La teoría de las inteligencias múltiples*. Segunda Edición. México: Fondo de Cultura Económica (1993).

Gardner, R. C. (2003). *Estadística para Psicología usando SPSS para Windows*. México: Prentice Hall.

Hines, J. M., Hungerford, H. R., & Tomera, A. N. (1987). Analysis and synthesis of research on responsible environmental behavior: A meta-analysis. *Journal of Environmental Education, 18*(2), 1–8.

Howard, A. L. (2013). Handbook of structural equation modeling. *Structural Equation Modeling: A Multidisciplinary Journal, 20*(2), 354–360.

Jolliffe, I. T. (1986). Principal component analysis. In D. Brillinger, S. Fienberg, J. Gani, J. Hartigan & K. Krickeberg (Advisors) *Springer series in statistics*. New York: Springer.

Jöreskog, K. G. (1978). Structural analysis of covariance and correlation matrices. *Psychometrika, 43*(4), 443–477.

Juárez-Nájera, M. (2010). *Sustainability in higher education. An explorative approach on sustainable behavior in two universities*. Doctoral Thesis. Faculty of Social Sciences, Erasmus University Rotterdam, The Netherlands.

Juárez-Nájera, M., Rivera-Martínez, J. G., & Hafkamp, W. A. (2010). An explorative socio-psychological model for determining sustainable behavior: Pilot study in German and Mexican Universities. *Journal of Cleaner Production, 18*, 686–694.

Ligtvoet, R., Van der Ark, L. A., Bergsma, W. P., & Sijtsma, K. (2011). Polytomous latent scales for the investigation of the ordering of items. *Psychometrika, 76*(2), 200–216.

McDonald, R. P. (2011). Measuring latent quantities. *Psychometrika, 76*(4), 511–536.

Muthén, L. K, & Muthén, B. O. (1998–2010). *Mplus user's guide* (6th ed.). Los Angeles: Muthén & Muthén.

Schmitt, T. A. (2011). Current methodological considerations in exploratory and confirmatory factor. *Journal of Psychoeducational Assessment, 29*, 304. doi:10.1177/0734282911406653.

Schwartz, S. H. (1994). Are there universal aspects in the structure and contents of human values? *Journal of Social Issues, 50*, 19–45.

Stern, P. C., Dietz, T., Abel, T., Guagnano, G. A., & Kalof, L. (1999). A value-belief-norm theory of support for social movements: The case of environmentalism. *Human Ecology Review, 6*, 81–97.

Welkowitz, J., Ewen, R. B., & Cohen, J. (2002). *Introductory statistics for the behavioral sciences* (5th ed.). NY: Wiley.

Part II
Actor's Scopes for Changing Peoples' Believes at HEI

Chapter 5
What to Promote for Achieving Education for Sustainability

The previous chapter assessed, in a confirmatory manner, the sustainable-behavior construct at different HEIs. FCA outcomes showed significant relationships among psychological indicators explain greater than 70 % of the explained variance. Similarities and differences between HEIs however could be explained by their status, the permeability of group borders, or group size or power (Sánchez 2002), and the method in which a questionnaire is administered should be determined by item content or theoretical approaches (Van de Vijver and Tanzer 1998).

Regardless of similarities and differences found among countries, the world situation requires educating critical, responsible, and fair citizens, and thus the DESD objectives may be achieved. In order to achieve such a citizenry, basic necessities must be adequately met: physiological needs, security, love, and belonging. Only when these needs are met may people realize themselves and attain a high level of self-esteem (Maslow 1958, 2005).

In order to explain the goal of education for sustainability, the first section of this chapter reviews the distinction between human needs and desires as a prerequisite for developing an ethical proposal which promotes such education among HEI. The second section presents some areas of human intervention where beliefs and attitudes may be changed to some extent in a long-term manner without coercion. These are education- and community-based areas.

This study shows that in the educational field, alternative learning methods such as game playing and art exploration may be integrated into the four main activities developed by higher education institutions—teaching, research, outreach, and physical campus operations. In the area of community management, group psychotherapy and labor management may modify individuals' potential for creativity, compassion, ethics, love, and spirituality. The goal is for individuals to find profound significance in their work relations in order to attain self-actualization. Table 5.1 summarizes a schema of principal HEI activities, the two areas of intervention mentioned, and four alternative learning methods.

© Springer International Publishing Switzerland 2015
M. Juárez-Nájera, *Exploring Sustainable Behavior Structure in Higher Education,*
Management and Industrial Engineering, DOI 10.1007/978-3-319-19393-9_5

Table 5.1 Four relevant learning methods in two human intervention areas among four university activities

Human intervention area HEI activities	Education	Community management
Teaching	Gaming (Vigotsky)	Group psychotherapy
Research		
Outreach	Art (Heidegger)	Labor management (Maslow)
Campus management		

5.1 What to Satisfy? Human Necessities or Human Desires

Human needs are dynamic notions. People can achieve them depending on prevailing conditions (Maslow 1958, 2005), or as Neuhouser (2008, 2014) suggests, by accidental conditions such as material dependence, inequality of wealth, division of labor, improved methods of production, and individual differences with respect to character circumstances and possessions related to luck, effort, and natural endowment.

Figure 5.1 shows Maslow's hierarchy of human needs divided into two main aspects according to Neuhouser (2008): self-preservation needs and recognition needs. Self-preservation needs include physiological needs, safety needs, and belongingness and love needs. Recognition needs include esteem needs and recognition needs in and of themselves.

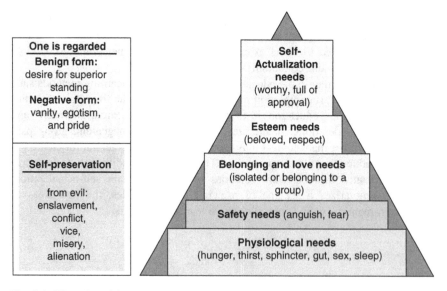

Fig. 5.1 Hierarchy of human needs (based on Maslow 1958, 2005, and complemented on Neuhouser 2008)

At the base of the pyramid are physiological needs; that is, basic needs such as hunger, thirst, sex, gut, and rest. It is quite true that "man lives by bread," but what happens to "man's" desires when there is plenty of bread and the belly is chronically full? Other (higher) needs emerge, and these, rather than physiological hungers, dominate the organism. The next most important class of motives includes safety needs (Maslow 1958; Maslow, quoted by Lowry 1973, p. 18). The need for safety is seen as an active and dominant mobilizer of the person's resources in emergencies such as war, disease, natural catastrophes, crime waves, societal disorganization, neurosis, brain injury, or chronically bad situations.

Once safety needs have been well satisfied, yet other needs emerge: the needs for belongingness and love, and the whole cycle will then repeat itself with a new motivation center (Lowry 1973; Maslow 1958). Now the person will keenly feel, as never before, the absence of friends, a sweetheart, spouse, or children. He or she will hunger for affectionate relations with people in general, for a place in the group, and will strive with great intensity to achieve this goal. The person will want to attain this more than anything in the world and may even forget that once, when hungry, he or she sneered at love as unreal, unnecessary, or unimportant.

Physiological needs and safety needs are normally fairly well satisfied in industrialized societies, whereas the needs of belonging and love, on the other hand, are not. This is so because love and affection, as well as their possible expressions in sexuality, are generally looked upon with ambivalence and we customarily follow many restrictions and inhibitions. Thus, in our society, the thwarting of these needs is the most commonly found core in cases of maladjustment and more severe psychopathology (Maslow 1958; Maslow, quoted by Lowry 1973, p. 26).

As the needs of belonging and love are satisfied, however, still another class of basic needs, the esteem needs, will emerge. This consists of the need for a stable, firmly based, high evaluation of oneself and therefore this need may be classified into two subsets. First is the desire for strength, achievement, adequacy, mastery and competence, confidence in the face of the world, and independence and freedom. Second, is what we may call the desire for reputation or prestige, status, dominance, recognition, attention, importance, or appreciation (Maslow 1958; Maslow, quoted by Lowry 1973, p. 27).

Even after all other more basic needs (physiological, safety, belonging-love, esteem) have been satisfied, we may still often (if not always) expect that a new restlessness will develop. An individual finds inner peace only when doing that for which he or she is fit. One must be what one can be. This need we may call self-actualization. It refers to a person's desire for self-fulfillment, namely the tendency for him or her to actualize their potentially. This tendency might be phrased as the desire to become more and more of what one is, to become everything that one is capable of becoming (Maslow 1958; Maslow, quoted by Lowry 1973, p. 26). However, when one is highly regarded by others, it can be in a benign form, such as being recognized for merit and honor, or in negative forms for pride, vanity, and egoism (Neuhouser 2008).

According to this theory, people are all good (self-actualizing), and decent inside, if only their basic needs are adequately fulfilled: their wishes for security, love, and esteem, not to mention the most basic, physiological needs (Maslow, quoted by Lowry 1973, p. 17). Rowan (1999) critiques the notion that Maslow's hierarchy of needs is a one-way linear trend from lower to higher levels. It introduces the idea of a contrast between abundance and deficiency motivation and uses this to suggest that the usual table of needs can be split into two vertically; see Table 5.2. At every level of activity, we can find deficiency motivation and abundance motivation side by side.

Nussbaum, in conjunction with the Nobel Prize winner in economics Amartya Sen, has proposed a reasonable and well-argued list of basic needs (Martínez 2000). Doyal and Gough (2003) have also published a list of basic needs which is having a great influence on reports prepared by the United Nations Development Program (UNDP). Both basic-needs lists are shown in Table 5.3.

The latter authors have thoroughly studied the possibility of a theory of human needs based on the firm conviction that such requirements are essentially the same for everyone, despite obvious biological and cultural differences that exist between people around the world. But it is clear that if one pursues progress toward an education for sustainable development or toward sustainable human development, generally speaking, it is necessary to distinguish needs from desires.

Table 5.4 provides a comparison between needs and desires. Necessities can be met because they are existential and physical, finite, few, classifiable, universal, and objective. On the other hand, wishes cannot be satisfied because they belong to future and are projections of the mind (Osho 2006a). However, core values and needs are relative and local, while economic resources and policies are global and universal. That is, needs are place- and time-specific across cultures (Gough 2004). The relationship between satisfying factors and needs is that of means to ends. But postmodern society is characterized, among other traits, by a deliberate and incessant confusion between ends and means (Martínez 2000). This implies that what may be satisfied is being neglected, and what cannot be fulfilled is fed (Osho 2006b). Humans are at a crossroads, and environmental and ethical implications are obvious.

Table 5.2 Motivational list of necessities (based on Rowan 1999)

Deficiency motivation: reactive (homeostatic in nature)	Abundance motivation: proactive (it is not reducible to a more complex kind of homeostasis)
Fearful	Outgoing
Lack	Overplus
Compulsive	Zestful
Predictable	Spontaneous
Closed	Open
Selfishness 1	Selfishness 2

Table 5.3 List of necessities (based on Martínez 2000; Gough 2003)

Proposed by Nussbaum and Amartya Sen	Proposed by Doyal and Gough, used by UNDP
1. Life. Capable of living a human life of normal length: not dying prematurely, or until one's life is so reduced as to not be worth living	Physical health
2. Bodily health. Capable of good health, including reproductive health; to be adequately nourished, to have adequate shelter	1. Nutritious food and clean water
3. Bodily integrity. The ability to move freely from place to place, having one's bodily boundaries treated as sovereign, i.e., being able to protect oneself against assault, including sexual assault, child sexual abuse, and domestic violence; having opportunities for sexual satisfaction and for choice in matters of reproduction	2. Protective housing
4. Senses, imagination, and thought. The ability to use the senses, to imagine, think, and reason in an informed manner cultivated by an adequate education, being able to use imagination and thought in connection with experiencing and producing self-expressive works and events of one's own choice: religious, literary, musical, etc. The ability to use one's mind in such a manner which is protected by guarantee of freedom of expression with respect to both political and artistic speech, and freedom of religious exercise. The ability to search for the ultimate meaning of life in one's own way. The ability to have pleasurable experiences, and to avoid unnecessary pain	3. A non-hazardous work environment
5. Emotions. The ability to have attachments to things and people outside ourselves; to love those who love and care for us; to grieve at their absence, in general: to love, grieve, experience longing and gratitude, and justify anger	4. A non-hazardous physical environment
6. Practical reasoning. The ability to form a conception of the good and to engage in critical reflection about the planning of one's life	5. Safe child bearing and birth control
7. Affiliation. The ability to live with and reach out to others, to recognize and show concern for other human beings, to engage in various forms of social interaction, to have the capability for both justice and friendship. The ability to be treated as a dignified being whose worth is equal to that of others	6. Appropriate healthcare
8. Other species. The ability to live with concern for, and in relation to, animals, plants, and the natural world	Autonomy
9. Play. The ability to laugh, play, and enjoy recreational activities	7. A secure childhood
10. Control over one's environment. The ability to participate effectively in political choices which govern one's life. The ability to demand property rights and seek employment on an equal basis with others	8. Significant primary relationships

Autonomy
7. A secure childhood
8. Significant primary relationships
9. Physical security
10. Economic security
11. Appropriate education

Table 5.4 Comparison between needs and desires (based on Osho 2006a)

Necessities	Desires
They can be satisfied	They cannot be satisfied
They are simple (hunger, thirst, sleep)	They are complex (to wish the symbolic value of an object or service)
They come from nature	They do not come from nature; they are creations of the mind
They come from the moment, creations of own life, existential, physical	They do not come from the moment, cannot be satisfied because their nature is a projection of the ego into the future. They are psychological
They are finite, few, classifiable, universal, and objective	They are infinite, diverse, unclassifiable, non-universal, and subjective

Necessities, roughly speaking, are not needs as such, but rather are instrumental satisfying factors dependent on local contexts. Tasks may be carried out in small groups, including formal education and social mobilization, in order to ethically intervene in current conflicts (Martínez 2000). But first and foremost, people's basic needs must be satisfied to promote responsible citizenship.

The next section discusses two areas of human intervention. The educational field and the community management area are spheres within which it is plausible to change human behavior in the long run without coercion. Four methods grounded in EFS principles—games, art, group psychotherapy, and labor management—are proposed for inclusion in HEI activities in order to transform people's personal and work relationships and find a deep significance in people's needs for self-actualization.

5.2 Spaces Where Beliefs and Human Behaviors May Be Modified

Political scientists believe that coordinating individual behavior for the common good is an eternal problem (Gardner and Stern 2002) and point out four basic areas (Stern 2000; Gardner and Stern 2002) in which behavior may be changed in a coordinated manner. The four areas identified are as follows:

(a) Religious and moral approaches which appeal to values and aim to change broad worldviews and beliefs;
(b) Education to change attitudes and provide information;
(c) Efforts to change the material incentive structure of behavior by providing monetary and other types of rewards or penalties; and
(d) Community management, involving the establishment of shared rules and expectations.

Actions involving combinations of these four areas of intervention could modify individual behavior in favor of the common good. However, moral- and incentive-based approaches both have generally disappointing track records and are coercive. Meanwhile, the community-based approach, which acts upon people's need for belonging, combined with education, may have potential to modify people's beliefs and attitudes to some extent without coercion in the long run (Stern 2000; Gardner and Stern 2002).

This section considers two of four areas identified above: that related to education including games and art, and the community management area including group psychotherapy and life experience as mechanisms which can modify human beliefs and attitudes to some extent.

5.2.1 Educational Area of Intervention

Behavioral achievements among individuals in HEI who have previously overcome internal barriers are quite specific, such as increasing their knowledge or degree of commitment. Education can make a difference in people's behavior, but there are serious limits to what may be accomplished. In the short term, education is only successful when principal barriers to action (for example, individual attitudes) are successfully modified. When such barriers are eliminated, individual actions, such as depositing cans in the recycling bin or adjusting the thermostat on the air conditioner, or even buying high-efficiency appliances, may be accomplished. Reducing external barriers requires greater effort—for example, community organizing or even changing national legislation. Education may have important indirect long-term effects, such as when education affects people's political preferences; this in turn influences government policy to reduce external barriers to sustainable behavior. Education is only likely to induce behavior which is already compatible with people's deeper values (Gardner and Stern 2002). Table 5.5 summarizes short- and long-term accomplishments, as well as some characteristics to overcome principal internal barriers to individual action.

Educational programs, according Gardner and Stern (2002), are more effective when they are designed according to psychological principles of communication and also directly address the links between attitudes and behavior. That is, making information available is not the same as to taking special effort to get people's attention, using sources of information which the audience trusts, involving the recipients of the information in efforts, reminding people that their pro-environmental attitudes apply to the situation at hand, and explaining how to manifest their attitudes.

Education works best when combined with other intervention strategies. For example, when an energy conservation program provided water-flow restrictors along with information on how to use them and how much water they could save behavioral success was achieved (Gardner and Stern 2002). Changing environmentally relevant behavior sometimes depends critically on the quality of the

Table 5.5 Accomplishments of education (based on Gardner and Stern 2002)

Factors	Accomplishments of education
Time	*Short-term educational strategies* These strategies are important source of information. They are effective, relatively simple, and involve little risk
	Long-term educational strategies These strategies can build public support for policies
	Long-term indirect effects Education can change people's political behavior; which in turn can change government policy
Characteristics	*Values* Education induces behavior compatible with people's deeper beliefs and values
	Efficiency Educational programs can be efficient when designed according to psychological principles of communication and when they directly address the links between attitudes and behavior
	Quality of information and level of public concern Changing relevant behavior depends mainly on the quality of the information and on the level of public concern

information provided and on the level of public concern and willingness to support the incentives or interventions.

The aim of education toward sustainability is to develop a way of life which includes all behavioral facets, where humans interact responsibly in their physical and social environments. Art and games, in the Gardner and Stern activities, are two ways of approaching this.

5.2.1.1 Play

The explosion of knowledge, combined with bureaucratization and increased division of labor, has produced highly trained, specialized experts. Frequently, specialists must process and absorb vast amounts of information in order to keep their jobs. They are simultaneously urged, as citizens, to develop a general understanding of world aspects. If a mission of HEI is to generate and transmit knowledge and technological advances, they must find methods of learning which combat narrow perspectives born of specialization, and integrate learning which leads to a competent, ethical judgment in order to understand what may be read within the structure of human experiences, and what describes and transmits complexity of our minds to others.

Some educators (Greenblat and Duke 1975) have identified critical elements to achieve such learning to include: (1) finding ways to instill motivation prior to transmission of information; (2) the learner being an active participant in the learning process, rather than a passive recipient of transmitted information; (3) individualized instruction which allows for each learner to proceed at their own pace; and (4) constructive feedback regarding success and error should be encouraged because there is a need for an awareness and understanding of elements and relations in a systematic manner.

Greenblat and Duke (1975) mention four heuristic principles for designing learning environments: First, the learner must have the opportunity to operate from several perspectives. Second, activities must include their own goals and sources of motivation, not only represent a mean to an end. Third, the learner must be encouraged not to depend on authority and allowed to reason for himself or herself; this will allow for a more productive in the learning process. And finally, the environment must be structured so as to respond positively to the learning activity, helping him or her to reflect and assess his or her own progress.

The importance of playing games lies in counteracting narrow perspectives derived from specialization and provides ways to develop a holistic understanding and the ability to retain details. Play is a tool for communication and learning (Greenblat and Duke 1975) and allow for simulating social situations based on certain explicit or implicit behavioral suppositions.

Figure 5.1 provides an outline of important principles of the Theory of Historical and Cultural Activity (THCA) developed by Vigotsky. THCA holds that each psychological function has a history of development which determines the level achieved in a higher psychological process (Morenza and Ruiz de 2004). The theory furthermore explains how games develop the learning process. In the human psyche, each higher psychological function exists at least twice, first in the social area as an inter-psychological function, and later in the individual area as an intra-psychological function. That is, the higher psychological function originates from interactions in the social communication process (Talyzina 1988).

Galperin and collaborators, or the so-called School of Vigotsky, poses a hypothetical mechanism explaining this process. The mechanism is called "internalization" (Morenza and Ruiz de 2004). When activities are internalized with external objects which act as socially defined symbols, not only is this symbol's image internalized, but also the entire structure of relations and transformations within the symbolic world is constructed. Tools—words, symbols, rituals—are used as aids in this process, but in "phase two," one learns to do this without the external tool. For example, we tour a new city; we initially need to use a map. But this later becomes unnecessary because an image of the city remains in our head (Vigotsky 1967).

Play is closer to recalling than to imagining; that is, it is memory in action rather than a new imaginary situation. As play develops, a movement occurs toward conscious awareness of its purpose. Play becomes an internal process, then internal speech, logical memory, and abstract thinking. A game is a source of development. According to Vigotsky (1985), Talyzina (1988), development is created in the "zone of proximal development" (ZPD). ZPD is the distance between the social and individual realm, namely between what individual is capable of doing without being prompted and what he or she is capable of if encouraged.

Vigotsky here identifies a measurement of development which the subject can achieve by collaborating with others. Vigotsky (1967) argues that *learning leads to development*; that is, if someone is being presented with challenges and also assisted in overcoming these challenges, they are induced to develop new skills. By contrast, Piaget argues that *development leads to learning*; that is, children can learn only what is possible given for his stage of development, which originates from an innate process of stages of development.

Play permeates attitudes toward reality. It has its own internal continuation at school and at work (compulsive activity based on rules). In play, action is subordinated to meaning, but in real life, of course, action is subordinated by meaning. All examinations of the essence of play have shown that play creates a new relationship between semantics and that which is visible—that is, between imagined situations and real situations.

Play can be seen as tools which can mediate between that which students do without any assistance, to that which they do through their relationship with others. Or, as Vigotsky proposes, it consists of concrete marks which initially act as an external aid and then are converted into structures in our mind, which can mediate between what students do without any kind of help and what they do through their relationship with others. Using play thus helps students to search for new ways to work together in an unsustainable world where ecological borders and complex ecosystemic processes are not currently respected. This requires the development of very inventive abilities, and a sustainable world requires not only collaboration, but also consideration in awaking the interest to develop: inquisitive attitudes, inductive reasoning, generation of ideas, new perspectives, and use of analogies (Juárez-Nájera et al. 2006).

Play is a source of development. Dieleman and Huisingh (2006) state, in their article on the potential of play in learning and teaching about sustainable development, that:

- Play generates <u>learning experiences and communication</u>. You can "learn by doing" without creating real consequences for the outside world.
- Playing games offers the possibility to <u>create shared experiences and form interpsychological relationships</u>. This is extremely important to arriving at shared definitions of problems and (visions) of solutions, which is crucial to in sustainable development. Sustainable development is a complex phenomenon which by its very nature involves a multitude of actors with a variety of backgrounds and positions regarding reality, and a key challenge is to develop a shared vision among such a heterogeneous group.
- Play contributes to <u>teambuilding</u> because it creates shared experiences. However, shared experience and teambuilding are related but different issues. Not every shared experience leads to a more positive experience of the other. Play which facilitate communication and collaboration usually results in better team performance and sense of group belonging. Here again the advantage of play is the "experimental" nature. Since it is "not for real," you may be able to induce individuals who prefer to be alone into collaboration.
- Play contributes to <u>knowledge of oneself or the formation of intra-psychological relationships</u>. Participants gain insight into their own thought processes. Play helps an individual discover one's implicit assumptions in life, which are not

necessarily shared by others. Play helps an individual perceive people's limitations and possibilities as part of a system. Participants learn that their freedom is bounded but that there is nevertheless room to move and influence the system. This can be very helpful in real life when we want to realize change.

- Play helps to test alternative solutions. As mentioned before, the real beauty of play is that we can "learn by doing" without negative consequences for the real world. We can simulate certain realities, play the games, manipulate reality, and experience consequences. While we test alternative solution, we learn things about ourselves and create shared experiences. With respect to sustainable development, such testing is essential; due to the systems-nature of sustainable development, it is very difficult to predict the outcome of interventions in the real world

- Last but not least, play is fun and entertaining; it is an idea that becomes an affect. Fun and entertainment are important because these generate energies and give the participants the energy to engage in the complex challenge that sustainable development confronts us with. It may also contribute to a change in the image that the concept of sustainable development still has. Many people associate the issue with words such as "heavy," "serious," "negative," or "depressing." But in fact, even though there is some truth that some of the qualifications, sustainable development is at the same time a space for creativity and adventure. Play may help to make people see this part of sustainable development.

Play helps students, faculties, administrators, and educational authorities alike to see that education for sustainability demands other lifestyles, forms of production, institutional organization, and research methods, and these can be simulated in the classroom, the laboratory, or within normal campus activities with no environmental impact, in a joyous, fun manner.

5.2.1.2 Art

Art, like science, is a diverse set of activities which allows one to explore, conform, construct, test, and challenge reality. Often, one considers only in terms of paintings and sculptures, poems and novels, music and dance, and plays to be art. Nevertheless, these lie within the process of self-questioning to understand the essence of reality and reflect that reality (Dieleman 2007a). Artists can make a real contribution to redefining reality, transcending boundaries of established institutional frameworks, and thinking in a lateral way (Dieleman 2007b), as the artistic process requires the concept of sustainability (a concept which redefines industrial development and material growth, incorporating bottom-up processes of decision-making and change.

In this study, the author uses Heiddeger's (2006), Gadamer (2002: chapter 9) explanation on esthetics in order to relate the concepts of art and sustainability in a different, as demanded by both themes. Why consider Heidegger? Because he, along with Wittgenstein, who come from different backgrounds, use different vocabularies, and have different concerns, coincide in denying the legitimacy of an ultimate fundamental philosophical searching (Bengoa 2002; Heidegger 2003; Moran 2011). This aspect seems important to the author of this study as forming part of a new paradigm applied to the study of art.

What is ultimate fundamental searching? Throughout history, philosophy has tried to elaborate a universal discourse on reality, from our knowledge of it, toward our actions with respect to it, that is, regarding principles—not only ontological, but also epistemological and ethical. This demand has been common to all philosophies which envision themselves as systems. These philosophies have always tried to justify their own exclusive access this universal knowledge.

However, in the decade of the 1930s, Kurt Gödel mathematically demonstrated that logical systems always contain wordings which are true, but that those systems cannot be derived from a fix set of axioms. That is, there is always missing information. During the third century before Christ, Aristotle expressed something similar when he stated that the sign of a well-educated mind is to be happy with the level of precision which the nature of the matter permits, and not seek accuracy when only an approximation is possible. At the beginning of the twentieth century, quantum mechanics discovered the uncertainty principle, complementarity, and wave-particle duality, thus showing limits to what we can observe with respect to microscopic events. At this level, quantum mechanics speculated that at this level exists an uninterrupted wholeness which cannot be separated into parts or events, which are basically statistical and undetermined, not exact (Briggs and Peat 1989). So, art according to Heidegger's philosophical position along with Gödel's thinking and the principals of quantum mechanics can be useful to explain how reality is perceived.

Heidegger (2003, 2006) holds that a work of art is an entity, which exists in a natural way, like an object. Works of art have thing elements and for centuries the thing has been taken as a model of the actual entity. There are three ways in which past thinkers have defined, described, and determined what a thing is: (a) the thing is a substance with accidents, (b) the thing is perceptible through sensation, and (c) the thing is formed by matter. However, Heidegger says that these are erroneous manners of relating the essence of the thing. These definitions of the thing do not adequately fit the essence of the thing—neither the essence of that which is useful nor the work of art.

Heidegger uses the phenomenological method (Moran 2011), which is a method that he adopts in his philosophical masterpiece, *Being and Time* (2003), to explain what a work of art is (Gadamer 2002). For him, there is nothing behind the phenomenon and to describe it, something (aletheia) come forth from concealment; beauty is one way in which truth occurs as unconcealedness (Heidegger 2006).

Heidegger (2006) discovers that the essence of that which is useful is rooted in its usefulness, which he calls "being of confidence," or when the useful thing is used—that is, when the useful thing makes apparent what in reality it is. This entity approaches the state of unconcealment of its being. Based on this conclusion, he establishes that in the work of art has set into operation the truth of that entity.

The existence of the work of art is due to the fact that such a work opens a realm, it creates a clearing. In that clearing truth, as unconcealment, can be encountered. Art exists only in that space, in that clearing (Heidegger 2006). The work of art is complete in itself, taken in isolation, but only within a set of relationships which transcends its particular entity to integrate it into the surrounding world. The work of art pre-exists to its appearance a set of beings but it is the work of art that enlightens beings and becomes the center which unifies them and constitutes them in a world.

The work of art illustrates a world not in the sense of the mere collection of the countable or uncountable, and familiar and unfamiliar things which are simply there; nor is it merely the imagine of framework added to our representation of the sum of such given objects. The world is the consciousness that turns on a light to tell beings to account for their existence and their positions in the midst of other existent beings; all things acquire their rhythm, their remoteness and closeness, and their breadth and narrowness. Beings become aware of their historical destiny, from their dependence on gods who can give or deny their grace. This world is not an abstract world but rather a way of intelligibility of all beings (Heidegger 2006).

Every work of art is made up of what is called raw materials, which are extracted from nature. By manifesting a world in the artistic work which causes the earth to be nature, the presentation sets up a world: rocks make a foundation, metal brings forth shine and sparkle, colors show up, sounds sound, and speech articulates. In other words, all those materials, through art, can bring forth the essence of beings from concealment. Admittedly, that which is useful is also made of matter, but subsequently disappears because what counts is the service. In addition, after using that which is useful, it suffers wear (Heidegger 2006).

Heidegger (2006) perceives that matter is not merely a "thingly foundation" of the work of art, but it within its full being its own value. He recognizes that in painting and sculpture, the brilliance of colors or the precious qualities of a marble carving, or sounds in music, or varied timbres of instruments are susceptible demonstrations of the essence of the internal constitution of the materials used in their production.

For Heidegger (2006), the creation of a beautiful work of art requires that the work sets up a world and an openness in which truths will emerge from concealment. The world and the earth struggle because they are antagonistic elements. The world patently is exposed to light; while the earth, in contrast, moves into the open, is self-secluding. In this struggle, there is something that tears a break in the deepest of earth, but it is precisely in this break where a gathering can be found. The world that is expressed in the work of art is no longer a requirement, but a specified content, a content of ideas, feelings, and projects which will make intelligible what is singular and concrete.

On the one hand, Liessmann (2006) holds that to a greater or lesser extent the philosophical approach identifies art with truth. The same idea applies to other philosophers such as Schelling, Hegel, and Schopenhauer. On the other hand, Heidegger (2006), Gordon and Gordon (2008) argues that truth is non-truth. But the truth exists only as the struggle between birth and concealment in the interaction between the world and the earth. The truth will be fixed in the work of art; the creation is nothing but truth fixed by form. Art is the truth of what it has set itself to work into. Indeed, the work of art itself retains its latent content until beings may stand back and relate with awe to it, become attuned to it; creations in art can be distinguished where a clearing emerges; an attunement to the work of art, of its radiant appearance. Heidegger's esthetics (2006) is very closely related to the sense of sustainability: simultaneously possessing and attracting a certain metaphysical extent.

Then, how may we stimulate, lead, or foster the process of change toward education for sustainability? If sustainability is a process of the creation of a new world with new institutions, products, processes, and relationships, and art is characterized as a search process that is not stuck in systematic scientific methodology (Dieleman 2007a), much room is left to associations, imagination, intuition, and mysticism, and as consequences, art transcends existing boundaries.

The sciences, field of action of HEI, are weakened due to analytical rationality which they apply in understanding reality. The process of change toward sustainability is "more than rational." It responds to desires, emotions, fears, lifestyles, identities, and intuitive notions. It lies in visions and future expectations or multiple futures. In essence, the change toward sustainability is the "art of being different," using different products, designing different lifestyles, engaging in different practices, doing things in different ways, and seeing reality in diverse forms (Dieleman 2007b).

Art is a powerful change agent; whenever it has been included in teaching and research activities, it has also produced effects on beliefs, habits, and values; even when students, faculties, or administrators developing art activities are attuned to art (as Heidegger states) with no purpose (Keeney 1994). This way, art can be executed to fit the demands in the principles of education for sustainability.

HEI can include elements and heuristic principles mentioned above to design learning environments where play and art take into account, once and for all and without prejudices, that they are not "serious" activities in higher education. To include them would respond to the demands of the principles of education for sustainable development (see Table 1.1).

5.2.2 Area of Community Management

According to the model developed in this study, moral norms play a decisive role in management of collective resources. That is, in the area of community management, group pressure is exerted through participatory processes and modification of individual behavior. Group psychotherapy and personnel management both offer examples of cases where individuals in a given community have been able to modify their behavior. Accordingly, if the management intervention area is applied toward a redefinition of the individual's role in industrial development, material gain, and social and cultural evolution to meet essential needs, then people may be guided toward sustainable behavior.

Gardner and Stern (2002) believe that a strong community, in psychological and sociological sense, is a group characterized by relative stability among its population, direct long-term social interactions, strong social networks, and a set of shared norms. These authors use the term "community management" to reflect the fact that administration within the group is much easier to organize and maintain if these four characteristics are met.

A key characteristic of community management is that social norms become shared rules, as fulfillment works upon a self-imposed rule that the participatory process develops from the bottom-up among group members, and because people believe that what they are doing is correct, or at least necessary. As the majority of people internalize community norms and make them their own, surveillance by authorities is minimal, and individuals do not feel coerced. Rules for interaction exist among group members that lead to informal social pressure and therefore self-control (London 1971).

Successful communities are those in which find easy and inexpensive ways to share information, enforce rules, and resolve conflicts swiftly and effectively, with appropriate, graduated penalties through a structure of incentives when sanctions are insufficient. In addition, accurate and relatively inexpensive systems assure that members comply with regulations. Authorities responsible for enforcing rules should be subject to control by users, so that they may be controlled or removed if they become corrupt or unjust. In organizational terms, keys for community management are participatory decision-making, monitoring, social norms, and sanctions throughout all community processes (Gardner and Stern 2002).

Gardner and Stern (2002) state that the success of community management of any social group ultimately depends on controlling behavior of individuals: How does a set of rules affect community management of individual behavior? What makes people follow rules when they can gain something by breaking them? The key is that most people do what is good for the group because they internalize the interests of the group, rather than acting out of compliance based on a set of external incentives.

People internalize group norms because they have participated in creating them, because they have seen their value for themselves and their community, and because norms have become part of community meaning by which sharing with others helps to maintain trusted relationships (Gardner and Stern 2002). Recalling the list of needs identified by Maslow (see Fig. 5.2), community members feel that their needs of belongingness and responsibility in the group have been met, their safety needs provided for by the group, and they have been allowed to achieve their needs for self-actualization. In the words of Maslow (2005, p. 17): the fulfillment of these needs may be the one main unconscious reason for projecting an inner problem into the outer world, i.e., just so that it can be worked on with less anxiety.

Gardner and Stern (2002) set out principles for intervention to change behavior: (1) use of multiple intervention types to address the factors limiting behavior change, (2) understanding the situations from the actor's perspective, (3) when limiting factors are psychological, applying understanding of the processes of human choice, (4) addressing conditions beyond the individual which constrain

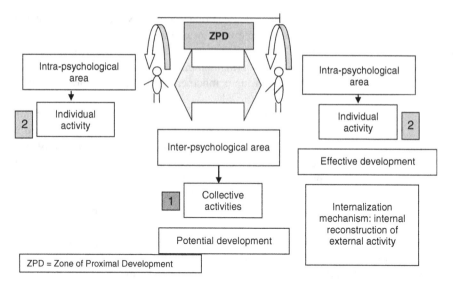

Fig. 5.2 Diagram of how human psyche functions according to the theory of cultural–historical activity, School of Vigotsky (based on Juárez-Nájera et al. 2006)

sustainable choice, (5) setting realistic expectations about outcomes, (6) continually monitoring responses and adjusting programs accordingly, (7) staying within the bounds of the actor's tolerance for intervention, and (8) using participatory methods in decision-making. Table 5.6 lists limiting factors for each of these principles.

Intervention to change beliefs and values also can come from therapeutic, self-help, and self-support groups, as well as from the group process inside labor organizations, since the great majority of people develop their daily activities at work, within a wide system of relations.

5.2.2.1 Group Psycho-therapy

In the community management area, groups, as social systems, perform an important role in both the interaction and integration processes that individuals have with institutions. Group psychology makes an unquestionable contribution to promotion, prevention, treatment, recovery, and intervention in the health realm. If health is understood according to the definition of the World Health Organization as a state of physical, psychological, and social well-being, and not only absence of illness (Sánchez 2002), so in turn, healthier groups also develop healthier systems of relations (Maslow 2005).

This leads to understand health as a state of social welfare, understood in socioeconomic terms as the part of the sociopolitical sphere which protects the interests and basic needs of individuals in society. In addition, welfare is defined as a component of the quality of community life which, along with economic and

Table 5.6 Principles of intervention to change behavior (based on Gardner and Stern 2002)

Principles	Limiting factors
1. Use of multiple intervention types to address the factors limiting behavior change, situation, and time	Technology, attitudes, knowledge, money, convenience, trust
2. Understanding the situations from actor's perspective	Scientific approach of control (pilot experiment) or participatory approach (social interaction with informal feedback)
3. When limiting factors are psychological, apply understanding of human choice processes	Commitment, credibility, face-to-face communication, conflict resolution, credibility, obligation, and norms
4. Address conditions beyond the individual that constrains sustainable choice	Incentive structure as indirect conditions
5. Set realistic expectations about outcomes	Trial and error method Experiences from other programs
6. Continually monitor responses and adjust programs accordingly	Flexibility and experimental interventions
7. Stay within the bounds of the actor's tolerance for intervention	Participation, education
8. Use participatory methods in decision-making	Participation, promotion of justice, internalization of new rules

psychological welfare, shapes the overall welfare of the community and individuals who are part of that community (Sánchez 2002).

That is, in healthier groups, systems of relationship are stronger or "healthier," or rephrasing Maslow (2005, p. 129): When we are healthy enough to see a higher unity, social synergy exists. Keeney (1994) speaks from the systems point of view, and of insanity: pathology arises when conscious and unconscious mental order is not connected to resources as part of a self-corrective feedback and any feeling, perception, or idea is always a fragment of the integral system or context in which it is found.

The promotion of social welfare in the health arena may be classified according to three approaches which appeared throughout the twentieth century and had a formative period between the years of 1903 and 1967, a second period of expansion between 1952 and 1967, and a third period of consolidation between 1968 and 1981 (Sánchez 2002). These guidelines are (1) therapeutic groups, (2) support groups, and (3) self-help groups.

First, the term therapeutic group includes those groups which adhere to the concept of clinical groups as a whole rather than psycho-therapeutically categories. These groups originally had teaching or educational purposes (e.g., groups in 1905 who gathered for information about tuberculosis hygiene or treatment) and subsequently spread to pathological situations or personal growth experiences.

Some examples of such groups are Bethel laboratory for social training or Lewin and Bradford T groups, Milan family therapy (Boscolo et al. 1987), personal growth groups, group analytical therapy, therapeutic communities (such as alcohol and drug addiction groups), etc. Such variety considers a range of group procedures, inspired by different psychological traditions such as Freud's classical psychoanalysis or humanistic psychology represented by gestalt therapy groups (Perls 2004), Rogers encounter groups, Berne's transactional analysis, Reich's bio-energy, Hellinger's systemic approach, or the cognitive method (Medin et al. 2005), and various forms of intervention aimed at individuals, relationships, or institutions (Sánchez 2002). Such groups enable individuals to get at the root of their experiences.

Over time, some groups evolved toward a more religious and transcendental dimension. The transcendental dimension followed two trends, one inspired in eastern religions such as Zen meditation (Watts 1989; Trungpa 1988; Osho 2006b) and yoga (De la Ferrier 1971) and another, more secular trend, inspired by Fromm's humanist socialism (1973, 2006).

The underlying therapeutic precept in self-help groups is that learning within groups produces more efficient results, because such experiences allow individuals to move deeply into their own experiences. These different approaches share a commonality, which is that the group may constitute a powerful instrument for intervention, learning, and change, thus improving people's quality of life (Sánchez 2002).

Second, according to Sánchez (2002), support groups aim to facilitate people's adaptation to circumstantial pressures which require them to manage new skills or and their psychosocial positioning. Common features of these groups are they are small and consist of volunteers; they meet regularly, sometimes under a professional supervision; they share experiences, strategies, coping skills, feedback, identification of resources, etc.; and their main objective is to provide mutual help toward achieving a particular purpose.

These groups are usually composed of people who share some kind of difficulties which alter or modify aspects of their normal functioning. Thus, the group provides these people new links and social relationships to compensate for their psychosocial deficiencies through interaction with people with the same problems, gaps, and/or common experiences. These groups include professionals responsible for initiating and controlling situations in order to facilitate people's adaptation to change.

These groups can be classified into subgroups: (a) those who either suffer a problem directly (widowed, divorced, diabetics, etc.) or indirectly (persons associated with those suffering from the problem); and (b) according to the type of problem (chronic, specific, or relating to changes of various kinds, such as legislative). The success of such groups will depend on the extent of self-management they are able to achieve.

The third category is the self-help group. According to Sánchez (2002), self-help groups are those who manage their own goals and ways of operating; therefore, they operate autonomously, independent of professionals and with no time limit. These promote group development based on social support. The different types of groups analyzed are valuable tools and strategies for intervention in social programs within community management to optimize people's psychosocial quality of life.

5.2.2.2 Labor Management

Organizations begin to realize that groups are fundamental units for carrying out a variety of productive activities. A more progressive vision is to move from the individual, as a unique entity, to the group. Group effectiveness is not only the final result obtained by its members, but also the process followed in order to obtain that outcome (Sánchez 2002). The group has widely been recognized as a social entity which performs a critical and fundamental role, because groups can influence the effectiveness and productivity of organizations in a great variety of ways. Currently, there is a unanimous agreement that groups are the cornerstone of modern organizations (Edersheim 2007) because in groups we grow up (families), work (organizations), learn (schools), decide (meetings), play (teams), and fight together (wars) (Sánchez 2002).

However, there has not been a constant interest in groups among organizations. Sánchez (2002) notes that in the early twentieth century, Frederick W. Taylor and his followers believed that groups were the enemies of efficiency at work, because they are potential hotbeds of organized resistance to efficient production.

At the end of the twenties, G. Elton Mayo, one of the first social psychologists along with collaborators initiated efforts to systematize the study of the role played by groups, emphasizing that working groups are a social context which strongly influences people's behavior.

The late thirties witnessed the birth of group dynamics by Lewin, a scholar of organizations at Bethel Laboratories. Lewin incorporated Mayo's conclusions. Mayo was a social psychologist who led theoretical and empirical research promoting increased incorporation of the applied field of human relationships in organizations. He revealed how the group can influence behavior, attitudes, and emotional states of people in the functioning of groups.

The end of the World War II witnesses two independent directions: one scholastic and other applied. Amid these trends emerges one which prefigures some features contained in more modern approaches, and in the context of psychology of organizations focused on socio-technical approaches of studying groups. From this perspective, the group is a social system and a social entity capable of achieving high levels of productivity. This perspective focuses on the distinction between activities related to production and social activities implicit to the functioning of working groups. It also highlights distinguishable goals which can be realized if organizational circumstances are suitable. The fundamental implications of this approach for working groups are the prescription of autonomy and self-regulation.

During the late fifties, the socio-technical approach generated a rich applied research on the importance of groups within organizations. At that time, Maslow (2005, p. 1), who gave up clinical psychology because he realized that individual psychotherapy was incapable of improving the situation of humanity, subsequently moved toward education as a way of reaching the entire human species. Maslow developed his "hierarchy of needs," developing a new branch of psychology, humanistic psychology, in the field of social psychology. He recognized that people in the process of transforming their relationships (self-actualizing) provide a better work environment for their colleagues and organizations. That is, proper management of the work lives of human beings—of the way in which they earn their living—can improve individual quality of life and improve the world.

Since the seventies, and particularly since the eighties with the creation of "quality circles" in Japan, among organizations there has been a progressive interest in many aspects of groups. Organizations are beginning to consider groups as fundamental units of organizational analysis which perform a wide range of productive activities, and the progressive vision centered on the individual as a basic unit begins to be replaced by the group. The reasons for this change of perspective are essentially practical rather than theoretical (Sánchez 2002).

This history offers two aspects: the characteristics of working groups and the main criteria of effectiveness used. Groups can be classified by (1) their level of formality or interrelation with the structure of the organization, (2) their temporary nature, (3) characteristics of the task based on interdependence or replicability of the goals, (4) their degree of autonomy from leadership by outside the group (self-directed or self-designed), and (5) external integration and internal differentiation from their environment. The book "Teachings by Peter Drucker" edited by Edersheim (2007) contains examples of existing organizations which demonstrate such characteristics.

As for group effectiveness, criteria are evaluated based on combined models (which support the idea that internal group processes are more important than the group environment in determining its effectiveness) and structural models (which assign a priority role to the group environment). Both models are based on the premise that group effectiveness is not only evidenced by the final result obtained by the group, but also by the process followed to arrive at such an outcome (Sánchez 2002).

The author of this study asks how HEI can foster behavior toward sustainability in people's everyday activities in the educational field and in the area of community management intervention to induce change in individuals? Apparently, Maslow's recommendation (2005, p. 51 and 52, paraphrased) more than 45 years after he wrote his book *Summer Notes on Social Psychology of Industry and Management* might be updated the path to financial and economic success requires that people adopt a long term, broad ranged, that they pay considerable attention to what we might call personal development, by proper training of managers and workers, that they pay attention to individuals' psychopathology, that they change organization environment show interest in and commitment to workers, and that they understand with complete clarity the objectives, directions, and goals of the organization. The utopian, psychological, ethical, and moral recommendations for this type of organization will improve all aspects of the situation.

In addition to that recommended by Maslow, Neuhouser (2008) establishes three characteristics of the frame of mind individuals must acquire in order to assume the standpoint of reason because when individuals reason are capable of acting correctly and rightly. First, reasoning requires one to step back from one's own particular desires and interests and to take an appropriately universal perspective (one that considers only the fundamental interest of each individual). Second, reason requires individuals to conceive of themselves as the moral equal of each of their associates, acting with the understanding that no one's individual interest has a higher claim than any one else's and that, for the purpose of forming laws, the fundamental interest of others take priority over their own individual interest. Finally, the individual must relinquish his claims to the ultimate authority of his own will and locate that authority in the opinions of others (in the prevailing consensus of his community subject to the appropriate constraints).

Among the lessons of labor management, these points mentioned can be applied to human economic life (Neuhouser 2014). Based on these lessons, Maslow (2005, p. xxii) poses three questions. Our respond may provide guidelines to initiate paths which lead human beings toward labor management by considering that human beings generally seek a job in order to live, and their live, and their job spreads and affects all spheres of their lives: (1) How good a society does human nature permit? (2) How good a human nature does society permit? (3) How good a society does the nature of society permit?…With this we hope to achieve the goal of influencing toward a sustainable behavior, with all implications that this entails. The fields of education and community management can modify peoples' behavior to achieve education for sustainability.

References

Bengoa, J. (2002). *De Heidegger a Habermas. Hermenéutica y fundamentación última en la filosofía contemporánea*. Barcelona: Herder.

Boscolo, L., Cecchin, G., Hoffman, L., & Penn, P. (1987). *Terapia familiar sistémica de Milán. Diálogos sobre teoría y práctica*. Buenos Aires: Amorrortu Editores.

Briggs, J., & Peat, F. D. (1989). *Espejo y reflejo: del caos al orden*. Gedisa: México.

De la Ferrier, S. R. (1971). *Yug, yoga, yoguismo*. Diana: Una matesis de psicología. México.

Dieleman, H. (2007a). Sustainability, art and reflexivity; why artist and designers may become key change agents in sustainability. In *Proceedings of the ESA Conference: New Frontiers in Arts Sociology*. Lüneburg and Hamburg, March 28–April 1, 2007.

Dieleman, H. (2007b). Science and art in society and sustainability; about a carpenter, a hammer and a chisel. Article published in the *webmagazine of Cultura21*, July 2007 (special issue on science and art in sustainability).

Dieleman, H., & Huisingh, D. (2006). The potentials of games in learning and teaching about sustainable development. *Journal of Cleaner Production, 14*(9–11), 837–847.

Edersheim, E. H. (2007). *Enseñanzas de Peter Drucker. Consejos finales del padre de la administración moderna*. México: McGraw-Hill.

Fromm, E. (1973). *The anatomy of human destructiveness*. New York: Holt Rineheart.

Fromm, E. (2006). *Psicoanálisis de la Sociedad Contemporánea*. México: Fondo de Cultura Económica.

Gadamer, H. G. (2002). *Los caminos de Heidegger*. Barcelona: Herder.

Gardner, O. T., & Stern, P. C. (2002). *Environmental problems and human behavior* (2nd ed.). NYC: Pearson Custom Publishing.

Gordon, R., & Gordon, H. (2008). *Hobbema and Heidegger. On truth and beauty*. New York: Peter Lang.

Gough, I. (2003). List and thresholds: Comparing the Doyal-Gough theory of human need with Nussbaum's capabilities approach. WeD working paper: WeD01. ESRC Research Group on Wellbeing in Developing Countries.

Gough, I. (2004). Human well-being and social structures: Relating to the universal and local. *Global Social Policy, 2004*(4), 289.

Greenblat, C., & Duke, R. D. (1975). *Gaming—simulation: Rationale, design, and applications. A text with parallel readings for social scientists, educators, and community workers.* New York: Sage publications.

Heidegger, M. (2003). *Ser y Tiempo.* Barcelona: Editorial Trotta.

Heidegger, M. (2006). *Arte y poesía (The origin of the Work of Art).* México: Fondo de Cultura Económica.

Juárez-Nájera, M., Dieleman, H., & Turpin-Marion, S. (2006). Games as tools for sustainability: The usage in education and in community outreach. In *Proceedings of the 4th International Conference on Environmental Management for Sustainable Universities*, June 26–30, at Stevens Point, Wisconsin, USA.

Keeney, B. P. (1994). *Estética del cambio.* Barcelona: Paidós.

Liessmann, K. P. (2006). *Filosofía del arte moderno.* Barcelona: Herder.

London, P. (1971). *Behavior control* (1st ed.). New York: Perennial Library.

Lowry, R. J. (1973). *A.H. Maslow: An intellectual portrait.* Monterey, California: Brooks/Cole Publishing Company.

Martínez, E. (2000). *Ética para el desarrollo de los pueblos.* Barcerlona: Editorial Trotta, S. A.

Maslow, A. (1958). Higher and lower needs. *Abridged from The Journal of Psychology, 1948*(25), 433–436.

Maslow, A. (2005). *El management según Maslow (Maslow on management).* Paidós: Una visión humanista para la empresa de hoy. Barcelona.

Medin, D. L., Ross, B. H., & Markman, A. B. (2005). *Cognitive psychology* (4th ed.). NY: John Wiley.

Moran, D. (2011). *Introducción a la Fenomenología.* Barcelona: Anthropos-UAMI.

Morenza, L., & Ruiz de, C. T. (2004). *Nuevas formas de enseñar y aprender. Guía para profesores.* Bolivia: Universidad Autónoma Gabriel René Moreno e Instituto Internacional para la Educación Superior en América Latina y el Caribe.

Neuhouser, F. (2008). *Rousseau's theodicy of self-love. Evil, rationality, and the drive for recognition.* New York: Oxford University Press.

Neuhouser, F. (2014). *Rousseau's critique of inequality: Reconstructing the second discourse.* Cambridge: Cambridge University Press.

Osho. (2006a). *Tao. Cuando el calzado es cómodo.* Madrid: Edaf.

Osho. (2006b). *Meditación. El arte de recordar quién eres.* Madrid, España: Edaf.

Perls, F. (2004). *El enfoque Guestáltico. Testimonios de terapia.* Barcelona: Cuatro Vientos Editorial.

Rowan, J. (1999). Ascent and descent in Maslow's Theory. *Journal of Humanistic Psychology, 39*, 125. doi:10.1177/0022167899393010.

Sánchez, J. C. (2002). *Psicología de los grupos.* Madrid: McGraw-Hill.

Stern, P. C. (2000). Toward a coherent theory of environmentally significant behavior. *Journal of Social Issues, 56*(3), 407–424.

Talyzina, N. (1988). *Psicología de la Educación.* Moscú: Ed. Progreso.

Trungpa, C. (1988). *The myth of freedom and the way of meditation.* Boston and London: Shambhala Dragon Editions.

Van de Vijver, F., & Tanzer, N. K. (1998). Bias and equivalence in cross-cultural assessment: An overview. *European Review of Applied Psychology, 47*(4), 263–279. (4th trimester).

Vigotsky, L. (1967). Play and its role in the mental development of the child. *Soviet Psychology, 5*(3), 6–18.

Vigotsky, L. (1985). Pensamiento y lenguaje. *Teoría del desarrollo cultural de las funciones psíquicas.* Moscú: Ediciones Quinto Sol. 1934.

Watts, A. (1989). *The way of Zen.* New York: Vintage books.

Chapter 6
Conclusions

6.1 Summary of Findings

At the outset of this study, it was stated that the main question of this investigation is to identify socio-psychological factors related to personality traits which can influence sustainable behavior of the individuals within higher educational institutions (HEI), as well as to present the areas where these individuals work, and in which higher education for sustainability is fostered. This study draws on social-psychology which is the scientific study of the reciprocal influence of the individual and his or her social context through the behavioral expression of his or her thoughts and feelings. Therefore, the research presented here addresses a range of contexts from intrapersonal processes and interpersonal relations to inter-group behavior and societal analyses.

The challenge is to devise ways to achieve socially desirable goals, such as the ones underlying the goals of the Decade of Education for Sustainable Development, while allowing people to recognize moral norms, through latent variables such as values, personal skills, ascription of responsibility, and awareness of consequences, as ways of explaining their behavior.

This investigation considers sustainable behavior to be "a set of effective, deliberate, and anticipated actions aimed at accepting responsibility for prevention, conservation, and preservation of physical and cultural resources. These resources include integrity of animal and plant species, as well as individual and social well-being, and safety of present and future human generations."

The theoretical framework of cognitivism was considered in order to explore the sustainable behavior construct, by using the approach of information processing in order to propose a social-psychology model using the existing models of attitude. Schwartz's moral norm-activation theory was the model selected, because it poses situations where social dilemmas are present, such as those faced by the education for sustainability. Schwartz's model is extended under the value-belief-norm by Stern et al., based on the very important aspect of Schwartz's set of universal

© Springer International Publishing Switzerland 2015
M. Juárez-Nájera, *Exploring Sustainable Behavior Structure in Higher Education*,
Management and Industrial Engineering, DOI 10.1007/978-3-319-19393-9_6

values. Also, the elements of Hines et al.'s meta-analysis were considered because of the importance of contextual variables. Interpersonal and intrapersonal intelligences from Howard Gardner's theory of multiple intelligences were also considered; these skills applied to any culture. Gardner's theory was analyzed through the psychological features of effectiveness, deliberation, anticipation, solidarity, and austerity as proposed by Corral-Verdugo and Pinheiro.

A questionnaire was prepared which consisted of 67 items in four sections according to the latent variable model. The first section of universal values includes 21 items of Schwartz's 10 value categories. At least, one item was included from each value type. Fifteen of the items supported principles underlying ESD and six items were contrary to ESD. The variables for moral norm activation from the second and third sections of the questionnaire were measured through nine items regarding awareness of consequences (AC) and nine regarding ascription of responsibility (AR). Those questions related to AC included importance to oneself, country, and other species of three actual environmental problems (climate change, loss of forests, and chemicals). In the AR section, three items concerned personal obligations, three concerned government obligations, and three concerned business obligations. The fourth section on intrapersonal and interpersonal intelligences contained 20 items, analyzed through five psychological dimensions of sustainability. The final section contained eight questions related to demographics such as age, gender, religious denomination, general income level, and educational training. Fifty-nine items were polytomous in four different Likert scale items, and 8 demographics were dichotomous.

The questionnaire was applied in two stages. The first stage was carried out in 2008, and the second one in 2013. For testing the hypothesis, four samples of participants were prepared summing all participants from the two stages. The so-called All-HEI sample includes all university participants (232) from five universities: (1) *Universidad Autonoma Metropolitana at Azcapotzalco* (UAMA) which is a public university located north of Mexico City; (2) *Leuphana Universität Lüneburg, Institut für Umweltkommunication* (LULIfUK) which is a public university 30 km from Hamburg in the Federal Republic of Germany; (3) *Université de Genève* (UdeG) a public university located in the city of Geneva, Switzerland's second largest university; (4) *Université de Montréal* (UdeM); and (5) *Université de Québec à Montréal* (UQAM) that are public francophone universities located in Montreal, Province of Quebec. The UAMA sample (127 participants) was called Lower Socio-Economic Level University (LSELU), a No-UAMA sample (105 participants) was called Higher Socio-Economic Level Universities (HSELU) from richer countries (Germany—40 participants, Switzerland—19 participants, and Canada—9 participants), and an outsiders sample (95 Mexican participants).

In order to test the proposed model verifying the reliability and degree of association among latent variables, two statistical procedures were applied in the following order: principal component analysis which explores the possibility of a factor structure underlying the latent variables and confirmatory factor analysis which deals specifically with measurement models, that is, the relationships between observed measures and latent variables.

Outcomes of this study, for the first time, explored five psychological dimensions toward sustainable action across five HEI in four countries even if the number of individuals is modest. Also, the emergent concept of sustainability is elucidated by building an instrument based on four current conceptual frameworks and DESD guides, PCA explanatorily revealed a general pattern for the main latent variables which underlie behavior for sustainability across HEI participants, and CFA exposed evidence in the latent structure of a second-order SB construct of four factors to explain the effect of categorical indicators.

The results of this research showed that the four latent variables are highly correlated and load a good factorability level according to the main PCA indicators of factorability as communalities, KMO and Bartlett tests, eigenvalues, and residual values from the matrix of reproduced correlations. The communalities values indicate how much variance within each variable is explained by the analysis; it was found that in all samples, an average was 0.7; this means that the latent variables had much in common with each other, that is, the associated variance from 66 to 77 % as opposed to 40–60 % from previous models.

The CFA results point out that the latent variable "ascription of responsibility," which refers to people's inclination to accept for the consequences of their behavioral choices toward the welfare of others, accounted for 84.8 % of the explained variance among LSELU members ($N = 127$) and 80.5 % for HSELU members ($N = 105$). For all university members, the R-square over SB accounted 72.5 % of the explained variance with $N = 232$ and for outsiders with $N = 95$, accounted 66.9 % of the explained variance. In all samples of participants, the latent variables "universal values," which represents conscious goals as response to needs of individuals, coordinated interaction, and smooth functioning and survival of groups, and "awareness of consequences," which refers to a person's receptivity for cues signaling situational needs, accounted less than 40 % of the explained variance. The same for the latent variable "inter- and intrapersonal intelligences," which showed psychological features concerned with the capacity to understand the intentions, motivations, and desires of others and oneself, is associated with sustainability; all samples of participants accounted less than 27 % of the explained variance. On the basis of the results of the overall goodness-of-fit indices, it can be concluded that the four-factor model evidence fits the data.

In order to develop critical, fair, responsible, and self-actualizing citizens, this study considers two areas of human intervention for changing behavior in the long run without coercion: education and community management. It also proposes four methods as alternative forms of learning and ways of strengthening group change—play, art, group psychotherapy, and personnel management—all grounded on the principles of EfS to be included in HEI activities.

In conclusion, this study showed that the model developed provides a real alternative for exploring social dilemmas in an exploratory and confirmatory manner in order to promote sustainable behavior among all decision makers in HEI. The model places a decisive importance on personal norms, which, if activated, are experienced among individuals as feelings of personal obligation, either denying or not denying the consequences of their behavioral choices regarding the welfare of others. The most viable areas of intervention for changing beliefs, attitudes, and values are education and community management because in these areas, people internalize their actions and no surveillance is needed. However, only long-term changes may be expected.

It is important to look at the nature of the conditions under which these different players in five HEI in four countries with very different socioeconomic contexts willingly foster within their organizational boundaries' concepts which promote a responsible society with a just and equitable development. Also, those outcomes induce behavior which has the potential to improve the exchange in organizational policies which are widely regarded as a useful tool for decision makers in a changing world for any type of society.

6.2 The Scientific Value and Practical Use of the Developed Model

The above findings from the social-psychology field allow us to statistically infer people's behavior within an organization. The model developed enables us to explain, measure, and predict people's sustainable behavior within HEI as well as to understand its factors. The framework provides a set of definitions to systematically search and construct a broader system which allowed us to test hypotheses regarding dependence upon specific factors as well as their importance in influencing willingness and behavioral change in five culturally different HEI. This model differs from others by taking into account the full range of universal values which are considered to be found throughout the inhabited world and human interpersonal and intrapersonal skills valid in all cultures.

Another contribution of this study is the integration into a developed model of the selected latent variables which further explain sustainable behavior. Hines et al.'s model provides a basic structure of contextual variables. Stern et al.'s theoretical framework provides theoretical and methodological bases for factors which trigger moral norms in behaviors which lead to social movements. Gardner's cognitive theory gives conceptual support to the theory that humans have evolved with several types of intelligence to treat different types of content in any cultural context, and Corral-Verdugo and Pinheiro's psychological dimensions suggest guides in people's intentions associated with sustainability.

Another final contribution of the model is its generalized nature which allows for widespread use of this model in different cultural contexts, as shown in Chap. 4 in five HEI. Outcomes across five HEI from four different countries added explanatory value of the model developed.

The introduction of this investigation pointed out that education for sustainable development is rooted in the extensive work of more than 30 years of environmental education (EE). Two common criticisms of EE are those emphasizing the study of the ecological dimension due to a widespread increase of environmental degradation and which restricted other human dimensions such as economic, social, political, and cultural dimensions, although these dimensions were included in its initial premises. This study recognizes the historical evolution of the concept of sustainable development and introduces clear principles which underlie education toward sustainability.

This study also differs from others on sustainable behavior, as it expands the definition of the term. Strangely, no author in the environmental psychology literature deals with the study of psychological factors which affect and are affected by the interaction between individuals and the environment, by offering a definition of sustainable behavior. The definition used in this investigation is broader, including aspects such as taking responsibility for prevention and conservation, not only preservation. It also considers the security of the individual and society, which was not present included in previous studies.

This study analyzes non-traditional practices such as the introduction of play and art, as well as psychotherapy groups and labor management, as real proposals by reviewing intervention areas where social dilemmas are presented in dealing with the common good. These alternatives, if conducted in a realistic and objective way, can produce long-term sustainable behavior, very much required in today's world.

6.3 Implications for Policy Design

The primary policy implication of this study influences the field of social-psychology in the analysis of sustainable behavior. This study proposes an instrument for analyzing the introduction of education for sustainability in HEI, allowing us to gain a better understanding and to assess possible sources of conflict among individuals in HEI when promoting sustainable behavior.

As shown in Chaps. 2, 3, and 5, this study can help to increase our understanding of the nature of the evolution of sustainable behavior in order to strengthen such behavior in among participating agents of higher educational institutions. The decision-making process can be handled well if participants (e.g. students, faculty members, and administrators) better understand the factors and potential areas for change.

In the process of promoting ESD principles mentioned in the introduction, through the intervention areas mentioned in Chap. 5, participants from HEI will have a better understanding of the importance of promoting sustainable behavior. The goal should be to increase awareness of social risks generated by current unsustainable behavior through activities relevant to sustainability such as those which may be provided by higher educational institutions.

A positive attitude toward changing behavior for sustainability should be found in modified perceptions of students, faculty, and administrators of HEI, though this does not occur regularly. We suggest that institutional policy explicitly promotes the social-psychological capability of generating synergies, promoting information exchange, and creating mechanisms which lead to a common vision with common goals among all types of participants. University members should aim to promote joint teaching, research, outreach, and campus management programs and projects which promote sustainable behavior. This would require linking activities to activities promoted by governmental authorities responsible for education and human development, as well as strengthening social recognition of the efforts of those institutions and individuals who show significant improvements in sustainable behavior.

Another way of using the results is for decision makers to be aware of these factors and how they interact among each other and to use this knowledge in the process of fostering sustainable behavioral change. In this respect, decision makers should be aware that the promotion of education toward sustainability is by no means under their direct control. Governmental authorities and civil society organizations must recognize indicators which allow for promoting behavioral change. This suggests that actors involved need to increase their social-psychology skills and, above all, their understanding of cultural change dynamics in HEI. Thus, they will be more capable of promoting behavioral change.

How to implement a program to produce behavioral change toward sustainability is beyond the scope of this study. Once sustainable behavior and its factors have been determined, the question of changing behavior toward sustainability remains open. The suggested initiatives in social-psychology knowledge capability-building are conceptualized as taking into consideration the minimization of the sources of conflict among the interests of the HEI players and the economic, environmental, social, political, and cultural answers which the world demands for current ethical and ecological situations.

6.4 Final Remarks

The general conclusion of this investigation is taken from Gardner and Stern (2002, p 342):

> Sustainability revolution will require profound changes in Western and non-Western institutions, economic process, values, morals. It will require changes in our basic conceptions of the relationship between humans and the rest of nature. It will require that we acknowledge the enormous complexity of global systems and our inability to manage them and mold them solely to the purpose of humans. And it will require that we more fully accept our responsibilities to future generations.

This implies that the ways to act in the world are, as Riechmann (quoted by Martinez 2000, p. 78, emphasis added) poses:

> In order to achieve a sustainable human development it is necessary *to stop spiral growth of unlimited material wealth aspirations, linked to consumption factors, and focus on an adequate coverage of universal necessities.* And that means *acting on the structure of needs, desires, and preferences* which are prevalent in our overdeveloped societies through a cultural revolution which is not clear if it is going to take place but which certainly is absolutely necessary for stopping ethical and ecological deterioration in which we find ourselves.

Appendix A

Complete list of universal values ranked in four categories according to Schwartz (2004)

Self-transcendence	Self-enhancement
Universalism	**Power**
1. Equality, equal opportunities for all	28. Social power, control over others, dominance
2. A world of peace, free of war and conflict	29. Health
3. Unity with nature, fitting into nature	30. Authority, the right to lead or command
4. Wisdom	31. Preserving public image
5. A world of beauty	**Achievement**
6. Social justice, correcting injustice, care of the weak	32. Ambitious, wealth, material possessions, money
7. Broad-minded	33. Influential, having an impact on people or events
8. Preventing and protecting pollution, conserving natural resources	34. Capable
Benevolence	35. Successful
9. Loyal, true friendship, faithful to friends	*Openness to change*
10. Honest, genuine, sincere	**Hedonism**
11. Amiable	36. Pleasure
12. Responsible	37. Enjoying life
13. Forgiving, willing to pardon others	38. Self-indulgent
Conservation	**Stimulation**
Tradition	39. An exciting life, stimulating experiences
14. Respecting the earth, harmony with other species	40. A varied life, filled with challenge, novelty, and change
15. Moderate	41. Daring
16. Humble	**Self-direction**
17. Accepting portion in life	42. Freedom
18. Devote	43. Creativity
Conformity	44. Independent

(continued)

© Springer International Publishing Switzerland 2015
M. Juárez-Nájera, *Exploring Sustainable Behavior Structure in Higher Education*,
Management and Industrial Engineering, DOI 10.1007/978-3-319-19393-9

(continued)

Self-transcendence	Self-enhancement
19. Politeness	45. Choosing own goals
20. Self-discipline, self-restrain, resistances to temptations	46. Curious, interested in everything, exploring
21. Honoring parents and elders, showing respect	
22. Obedient, dutiful, meeting obligations	
Security	
23. Social order	
24. National security	
25. Reciprocation of favors	
26. Family security, safety for loved ones	
27. Clean	

Appendix B

Complete list of interpersonal and intrapersonal intelligences ranked in four categories according to Boyatzis et al. (2002)

Self-knowledge	Understanding of other
1. Emotional awareness	10. Empathy
(a) Is aware of own feelings	(a) Listens attentively
(b) Recognizes the situations that arouse strong emotions in him/her	(b) Is attentive to peoples' moods or nonverbal cues
(c) Knows how his/her feelings affect his/her actions	(c) Relates well to people of diverse backgrounds
(d) Reflects on underlying reasons for feelings	(d) Can see things from someone else's perspective
2. Precise self-knowledge	11. Organizational awareness
(a) Acknowledges own strengths and weaknesses	(a) Understands informal structure in the organization
(b) Is defensive when receiving feedback	(b) Understands the organization's unspoken rules
(c) Has a sense of humor about oneself	(c) Is not politically savvy at work
(d) Anticipates obstacles to a goal	(d) Understands historical reasons for organizational issues
3. Self-confidence	12. Orientation of service
(a) Believes oneself to be capable for a job	(a) Makes self available to customers or clients
(b) Doubts his/her own ability	(b) Monitors customer or client satisfaction
(c) Presents self in an assured manner	(c) Takes personal responsibility for meeting customer needs
(d) Has "presence"	(d) Matches customer or client needs to services or products
Self-determination	*Social skills*
4. Emotional self-control	13. Developing staff
(a) Acts impulsively	(a) Recognizes specific strengths of others
(b) Gets impatient or shows frustration	(b) Gives directions or demonstrations to develop someone

(continued)

© Springer International Publishing Switzerland 2015

M. Juárez-Nájera, *Exploring Sustainable Behavior Structure in Higher Education*,
Management and Industrial Engineering, DOI 10.1007/978-3-319-19393-9

(continued)

Self-knowledge	Understanding of other
(c) Behaves calmly in stressful situations	(c) Gives constructive feedback
(d) Stays composed and positive, even in trying moments	(d) Provides ongoing mentoring or coaching
5. Integrity	14. Leadership
(a) Keeps his/her promises	(a) Leads by example
(b) Brings up ethical concerns?	(b) Makes work exciting
(c) Acknowledges mistakes	(c) Inspires people
(d) Acts on own values even when there is a personal cost?	(d) Articulates a compelling vision
6. Adaptable	15. Change catalyze
(a) Adapts ideas based on new information	(a) States need for change
(b) Applies standard procedures flexible	(b) Is reluctant to change or make changes
(c) Handles unexpected demands well	(c) Personally leads change initiatives
(d) Changes overall strategy, goals, or projects to fit the situation	(d) Advocates change despite opposition
7. Orientation to achievement	16. Influence
(a) Seeks ways to improve performance	(a) Engages an audience when presenting
(b) Sets measurable and challenging goals	(b) Persuades by appealing to peoples' self-interest
(c) Anticipates obstacles to a goal	(c) Gets support from key people
(d) Takes calculated risks to reach a goal	(d) Develops behind-the-scenes support
8. Initiative	17. Conflict handling
(a) Hesitates to act on opportunities	(a) Airs disagreements or conflicts
(b) Seeks information in unusual ways	(b) Publicly states everyone's position to those involved in a conflict
(c) Cuts through red tape or bends rules when necessary	(c) Avoids conflicts
(d) Initiates actions to create possibilities	(d) In a conflict, finds a position everyone can endorse
9. Optimism	18. Team work and cooperation
(a) Has mainly positive expectations	(a) Does not cooperate with others
(b) Believes the future will be better than the past	(b) Solicits others' input
(c) Stays positive despite setbacks	(c) In a group, encourages others' participation
(d) Learns from setbacks	(d) Establishes and maintains close relationships at work

Appendix C
Applied Questionnaire

C.1 Questionnaire on Sustainable Behavior

The purpose of this questionnaire is to gather information about factors that influence the sustainable behavior of key actors in higher educational institutions to foster sustainable development concept in their teaching, research, extension, and campus management activities. Sustainable behavior is measured in this questionnaire on individual basis; however, it is shown by statistics a collective attitude.

The sustainable behavior is evaluated in this questionnaire according to the following definition: *"the set of effective, deliberate, and expected actions addressed to accept responsibility for prevention, conservation, and preservation of physical and cultural resources that include integrity of animal and plant species, as well as individual and social well being and material safety of actual and future human generations."*

It should take about 12 min to complete this questionnaire. Any information you provide will be kept **strictly confidential** and **will only be used** for the purpose mentioned herein. Our interest for your responses is purely scientific.

- Please fill the questionnaire in by yourself and do not argue any answer with anyone else.
- Besides, do not think too much when you are answering. Try to respond spontaneously.
- There are no true or false answers. The most important is what you think.
- Please answer every item, even if you think they are repeated over and over again.

In case you have any question, at the end of the questionnaire, there is a space to express it. If you want to send it back, please mail it to Margarita Juárez-Nájera: mjn@correo.azc.uam.mx

© Springer International Publishing Switzerland 2015
M. Juárez-Nájera, *Exploring Sustainable Behavior Structure in Higher Education,*
Management and Industrial Engineering, DOI 10.1007/978-3-319-19393-9

Please start fill in the survey now!

Next, there is a list of concepts or words. We like your opinion if you identify them; rate according to the following scale at the extent to which you are strongly agreed or strongly disagreed on each item. The neutral option "I-am-not-decided" is included.

1. means **strongly agree** (SA)
2. means **somewhat agree** (A)
3. means **undecided** (U)
4. means **somewhat disagree** (D)
5. means **strongly disagree** (SD)

Please fill in with a "X" mark on the corresponding option. Remember you can mark whatever column

Concepts, words	1 SA	2 A	3 U	4 D	5 SD
1.1. World at peace, free of war and conflict					
1.2. Influential, having an impact on people and events					
1.3. Ambitious, wealth, material possessions, money					
1.4. Broad-minded					
1.5. Authority, the right to lead or command					
1.6. Creativity					
1.7. Social power, control over others, dominance					
1.8. Social order					
1.9. Preventing and protecting the environment, conserving natural resources					
1.10. Varied life, filled with challenge, novelty, and change					
1.11. Social justice, correcting injustice, care of the weak					
1.12. Enjoying life					
1.13. Self-discipline, self-restraint, resistance to temptations					
1.14. Unity with nature, fitting into nature					
1.15. Wealth					
1.16. Responsible					
1.17. Respectful, respecting the earth, harmony with other species					
1.18. Moderate					
1.19. Equality, equal opportunity for all					
1.20. Accepting one's portion of life					
1.21. Choosing own goals					

In the next block, we like to rate three problems raised.

1. **very serious** problem
2. **somewhat serious** problem
3. **no serious** problem at all

Please fill in with a "X" mark on the corresponding option. Remember you can mark whatever column

Statement	1 Very serious	2 Somewhat serious	3 No serious
2.1a. In general, do you think that climate change, which is sometimes called the greenhouse effect, will be a problem for you and your family?			
2.1b. Do you think that climate change will be a problem for the country as a whole?			
2.1c. Do you think that climate change will be a problem for other species of plants and animals?			
2.2a. Next, I would like you to consider the problem of loss of tropical forests. Do you think this will be a problem for you and your family?			
2.2b. Do you think that loss of tropical forests will be a problem for the country as a whole?			
2.2c. Do you think that loss of tropical forests will be a problem for other species of plants and animals?			
2.3a. Next, I would like you to consider the problem of toxic substances in the air, water, and soil. Do you think this will be a problem for you and your family?			
2.3b. Do you think the problem of toxic substances in the air, water, and soil will be a problem for the country as a whole?			
2.3c. Do you think the problem of toxic substances in the air, water, and soil will be a problem for other species of plants and animals?			

In the next section, we like to rate some statements. Please rate according to the following scale at the extent to which you are strongly agreed or strongly disagreed in each item. The neutral option "I-am-not-decided" is included.

1. means **strongly agree** (SA)
2. means **somewhat agree** (A)
3. means **undecided** (U)
4. means **somewhat disagree** (D)
5. means **strongly disagree** (SD)

Please fill in with a "X" mark on the corresponding option. Remember you can mark whatever column

Statement	1 SA	2 A	3 U	4 D	5 SD
3.1. The government should take stronger action to clean up toxic substances in the environment					
3.2. I feel a personal obligation to do whatever I can to prevent climate change					

(continued)

(continued)

Statement	1 SA	2 A	3 U	4 D	5 SD
3.3. I feel a sense of personal obligation to take action to stop the disposal of toxic substances in the air, water, and soil					
3.4. Business and industry should reduce their emissions to help prevent climate change					
3.5. The government should exert pressure internationally to preserve the tropical forests					
3.6. The government should take strong action to reduce emissions and prevent global climate change					
3.7. Companies that import products from the tropics have a responsibility to prevent destruction of the forests in those countries					
3.8. People like me should do whatever we can to prevent the loss of tropical forests					
3.9. The chemical industry should clean up the toxic waste products it has emitted into the environment					

Next, there is a list of actions. Please rate according to the following scale at the extent to which you never do or consistently do it, in each affirmative item.

1. means **never** (N)
2. means **rarely** (R)
3. means **sometimes** (S)
4. means **often** (O)
5. means **consistently** (C)

Fill in with a "X" mark on the corresponding option. Remember you can mark whatever column

Actions	1 N	2 R	3 S	4 O	5 C
4.1. Anticipate obstacles to a goal					
4.2. Adapt ideas based on new information					
4.3. Solicit others' input					
4.4. Take calculated risks to reach a goal					
4.5. Relate well to people of diverse backgrounds					
4.6. Stay composed and positive, even in stressful situations					
4.7. Lead by example					
4.8. Advocate change despite opposition					
4.9. Get impatient or show frustration					
4.10. Personally lead change initiatives					
4.11. Keep your promises					

(continued)

(continued)

Actions	1 N	2 R	3 S	4 O	5 C
4.12. Acknowledge mistakes					
4.13. Articulate a compelling vision					
4.14. Can see things from someone else's perspective					
4.15. Believe yourself to be capable for a job					
4.16. Cut through red tape or bend rules when necessary					
4.17. Doubt own ability					
4.18. Establish and maintain close relationships at work					
4.19. Hesitate to act on opportunities					
4.20. Change overall strategy, goals, or projects to fit the situation					

Finally, we include a set of personal questions for statistical purpose.

Please fill the survey in and mark with an "X" on the corresponding option according to your actual situation.

5.1. What level of studies have you obtained? Please mark the highest obtained!

		Only one answer
1.	Incomplete secondary studies	
2.	Complete secondary studies	
3.	Incomplete higher educational studies	
4.	Complete higher educational studies	
5.	Complete postgraduate studies	
6.	None	

5.2 What kind of housing do you have? (Mark only one)
( ) House   ( ) flat/apartment

5.3 Do you own your house/apartment?
( ) I am owner ( ) Rent

5.4 Under what religious denomination were you born? (Choose only one option)
( ) Roman Catholic   ( ) Lutheran Protestant   ( ) Calvinist Protestant
( ) Jews   ( ) None/Atheist   ( ) Other, clarify:_____

5.5 Sex: ( ) Masculine ( ) Feminine

5.6 Year of birth: _____

5.7 Are you: (If you have more than one activity, please choose the most important one and tell at the ending section the other one). Only one answer!

Student ( )   Faculty ( )   Administrator ( )

5.8 What is your Higher Educational Institution:
LULIfUK ( )   UAMA ( )   UdeM/UQAM ( )   UdeG ( )   Outsider ( )

Thank you very much for your collaboration!

If you want to express any opinion or comment, use these lines.

Appendix D
Statistical Techniques for Testing the Model

Two statistical techniques are applied to validate the model developed to analyze decision makers at HEI: the principal component analysis (PCA) and the confirmatory factor analysis. This appendix provides a brief historical description of each method, their scope, and their mathematical expressions. Also, their application and further interpretation are provided. The language used in the mathematical description for each procedure is less rigorous than it would be for statisticians or engineers. For more details, see Bartholomew (1987), and Basilevsky (1994), Brown (2006), and Jolliffe (1986).

D.1 Principal Component Analysis (PCA)

PCA is probably the oldest and most well-known multivariate analysis technique (Jolliffe 1986). It has its origins in 1889 when Galton devised the concept of the variable or latent trait to explain the relationship between measured variables, but Pearson in 1901 extended the Galton's concept of regression by developing correlation measurements (Gardner 2003). However, Spearman in 1904 (Basilevsky 1994) developed the first model of common factors in the context of the psychological "general intelligence" test and subsequently introduced the term "factor."

Spearman used the concept to support his assertion that measurements were composed of two factors, an overall capacity common to all measurements and a set of specific skills for each measurement. Other researchers, such as Thomson (1956), have disagreed with this concept and argue that only groups of common factors exist. A third group of researchers alternatively suggested the existence of a hierarchy of capabilities from general to specific. Many of these developments and discussions focused on Great Britain (Gardner 2003).

In the USA in 1947, Thurstone presented arguments against the concept of a common factor and in favor of the concept of multiple factors which he called primary mental abilities (Bartholomew 1987). He introduced the simple structure concept and suggested that any given factor must be defined primarily by a subset of non-transplant variables. Also, he proposed turning factors to discover such simple structures. Subsequently, other researchers such as Hotelling and Girshick

© Springer International Publishing Switzerland 2015
M. Juárez-Nájera, *Exploring Sustainable Behavior Structure in Higher Education*,
Management and Industrial Engineering, DOI 10.1007/978-3-319-19393-9

proposed analytical procedures capable of identifying the simple structure, but it was not until the advent of computers that their use spread widely. Nowadays, PCA is rooted in virtually every statistical package (Jolliffe 1986).

The term "principal component analysis" is a common term in statistical literature and is adopted in this research. We do not use other phrases such as "empirical orthogonal functions" or "factor analysis" which are confusing, or "eigenvector analysis" or "latent vector analysis" that camouflages PCA. There are several procedures for PCA. All procedures have many things in common but differ in the nature of the mathematics employed. In addition, all methods tend to respond very similarly in terms of the underlying dimensionality of any given set of variables (Jolliffe 1986; Gardner 2003).

The central idea on PCA is to reduce the dimensionality in a set of data in which there is a great number of interrelated variables, while retaining as much as possible the actual variation of the entire set of data. This reduction is achieved by transforming the entire set of data into a new set of variables, *principal components*, which are not correlated and are ranked by the first few variables, thus maintaining the majority of the variation present in *all* original variables. PCA reduces the solution of a problem of eigenvalues (own values) to eigenvectors for a symmetric, semidefined, positive matrix. Thus, the definition and calculations of principal components are direct, apparently simple, and have an ample variety in many applications (Jolliffe 1986). PCA does not address the manner in which some variables influence the construct, but rather deals with variable relationships among variables. It also fails to determine factor significance, but the factor explains the percentage of total variance[1] and also how highly this variable is related to such factor (Gardner 2003).

PCA consists of three stages (Gardner 2003). Stage 1 calculates relationships among variables. This is usually expressed as a *correlation matrix*. PCA "extracts" the matrix dimensions. In this context, the fundamental theorem of PCA is that correlation between any two variables can be expressed as the sum, in all dimensions, of cross-correlation products between these two variables and dimensions. The theorem is expressed as follows (for a more rigorous statistical and mathematical language, see Basilevsky 1994, Chap. 6):

$$r_{XY} = \alpha_{XJ}\alpha_{YJ} + \alpha_{XJI}\alpha_{YJI} + \cdots + \alpha_{XK}\alpha_{YK} \qquad (D.1)$$

where

α_{XJ} is the correlation between X variable and I factor (first dimension)
α_{YJ} is the correlation between Y variable and I factor, etc.
α is also known as factorial saturations.

---

[1] Variance is a scale or dispersion statistics. It is the square means of the deviation. It is usually denoted with S^2. The formula is as follows: $S^2 = [\Sigma(X - Xbar)^2]/n$.

Stage 2 extracts factors. These factors are dimensions on mathematical bases which describe principal components from the variance in the correlation matrix; the correlation matrix across these variables and dimensions constitute the *matrix of initial factors*. The dimensions in themselves are nothing more-than-aggregated weighted variables expressed as standard scores. In other words, an individual score is a factor, expressed as a standard score. It is as follows:

$$F_i = w_1 Z_{1i} + w_2 Z_{2j} + \cdots + w_m Z_{mi} \qquad (D.2)$$

where

F_i is an individual score in the factor.
w_1, w_2, etc., are weights
Z_{1i}, Z_{2j}, etc., are individual standard score in variables.

The principal component method uses matrix algebra to select the weights for each factor, so that the variance of scores for the factor is as great as possible. The scale of these weights is adjusted to the sum of their square values equal to 1.0. The set of weights (w) for each factor is called eigenvector (or own vector). A matrix of these weights would have as many columns and rows as factors and variables. All factors are independent of each other due to a process of partitioning correlations between each variable and individual factor. This could be demonstrated by calculating a score for each factor in each individual and correlating their scores. All resulting correlations between factors would be 0 (zero). The variance of each factor could also be calculated. That is, factor variances are called *eigenvalues* or own values.

A highly positive correlation between a variable and a factor indicates that the variable tends to measure something in common with that factor. A highly negative correlation indicates that the variable tends to measure the opposite of what is described by the factor. A very low correlation indicates that the variable has nothing in common with the factor.

If the factorial saturation (a) is squared for each variable and the sum of resulting values for each factor is obtained, the eigenvalues are subsequently obtained. These eigenvalues are the variance for each factor, and can be calculated with the following equation:

$$\lambda = \alpha_1^2 + \alpha_2^2 + \cdots + \alpha_m^2 \qquad (D.3)$$

where

λ = eigenvalue for a factor
α_1^2 = Factorial square weight for variable 1 for the factor; α_2^2 factorial square weight for variable 2; and so on for m variables

If the sum of squares of factorial saturation is obtained for all factors for each variable, resulting values would be *commonalities* for each variable. Commonality, designated by h^2, for a variable is a measurement of how much variance that variable has in common with all other variables in the matrix, at least with respect to factors which were extracted. The formula for calculating them is as follows:

$$h^2 = \alpha_1^2 + \alpha_2^2 + \cdots + \alpha_p^2 \qquad (D.4)$$

where

h^2 = variable commonality.

α_1^2 = square of factorial weight for variable 1 for the factor; α_2^2 square of factorial weight for variable 2; and so on for p factors.

Stage 3 identifies factors which describe, in the simplest possible manner, relationships among variables. A problem of PCA is that factors are extracted in order to explain variance, not in terms of how well they really describe the relationship among variables. This is achieved by rotating factors described in the initial factor matrix to produce an initial structure more susceptible to interpretation. The resulting matrix of associations among variables and rotated factors is the *matrix of rotated factors*.

It is possible to use many types of rotating procedures (Jolliffe 1986), but the nature of the solution they produce may vary. Solutions can be orthogonal or oblique. In the orthogonal solutions, rotated factors retain independence characterized by PCA. The oblique solutions, on the contrary, allow for a correlation between factors. This implies that interpretations of factors may have some overlap, depending on the extent of the correlation between two given factors.

As in all multivariate procedures, PCA has several aspects to consider: sample size and number of factors. Traditionally, it has been argued that samples must contain at least 100 and 300 measurements. However, a Monte Carlo study of rotated principal components found that the most important feature influencing the stability of the results is factor saturation (Gardner 2003). If those factors are well defined (i.e., if the saturation of either is large in 0.8 in the population), a sample size as small as 50 is relatively stable (Basilevsky 1994). In any case, sample sizes of 100 to 200 are more adequate.

A primary objective of PCA is to reduce the number of dimensions needed to describe relationship among variables (Jolliffe 1986; Bartholomew 1987; Basilevsky 1994; Gardner 2003). In general, there are two ways of doing this, criteria of eigenvalue of one and proof of sedimentation. The first approach assumes that all factors with an eigenvalue greater than 1.0 mean something significant and must be retained in the final solution. The second approach involves drawing a chart of eigenvalues against factors arranged in the order of 1 to m. This graph has been compared with a side view of mountains, and the problem is to determine where the mountain ends and where the ground level at the base of the mountain begins. Gardner (2003) recommends basing his decisions on first "rocks," that is, maintaining all factors prior to the first "elbow."

Both of these procedures seek to determine at what point virtually any association among variables has already been explained and to decide when the remaining association basically reflects sampling fluctuations. This can be determined directly considering the residual matrix once the factors have been extracted. A combination of these procedures is recommended by many scholars (Gardner 2003).

The aforementioned procedure is an exploratory method. In other words, it is part of an association matrix (for instance, correlations) and attempts to identify factors underlying such association. That is, the intention is to find factors which are responsible for partnerships, not to test the extent of adequacy (confirmatory factor analysis). If one wishes to achieve the latter, more previous explanatory information must be available.

D.2 Confirmatory Factor Analysis (CFA)

The central idea on CFA is to identify latent factors that account for the variation and covariation among a set of indicators; CFA requires a strong empirical or conceptual foundation to guide the specification and evaluation of the factor model (Brown 2006). CFA is based on the common factor model.

A fundamental equation of the common factor model is as follows:

$$y_j = \lambda_{j1}\eta_1 + \lambda_{j2}\eta_2 + \cdots + \lambda_{jm}\eta_m + \varepsilon_j \tag{D.5}$$

where y_j represents the jth of p indicators or items in a questionnaire obtained from a sample of n independent subjects; λ_{jm} represents the factor loading-relating variable j to the mth latent factor η; and ε_j represents the variance that is unique to indicator y_j and is independent of all ηs and all other εs.

The regression functions can be summarized by separate equations according to the number of observed indicators ($Y1, Y2,..., Yn$):

$$
\begin{aligned}
Y1 &= \lambda_{11}\eta_1 + \varepsilon_1 \\
Y2 &= \lambda_{21}\eta_1 + \varepsilon_2 \\
Y3 &= \lambda_{31}\eta_1 + \varepsilon_3 \\
Y4 &= \lambda_{41}\eta_1 + \varepsilon_4 \\
Y5 &= \lambda_{51}\eta_1 + \varepsilon_5
\end{aligned}
\tag{D.6}
$$

This set of equations can be summarized in a single equation that express the relationships among observed variables (y), latent factors (η), and unique variances (ε):

$$y = \Lambda_y \eta + \varepsilon, \tag{D.7}$$

or in the expanded matrix form:

$$\Sigma = \Lambda_y \Psi \Lambda_y' + \Theta_\varepsilon \tag{D.8}$$

where Σ is the $p \times p$ symmetric correlation matrix of p indicators, Λ_y is the $p \times m$ matrix of factor loadings λ, Ψ is the $m \times m$ symmetric correlation matrix of factor correlations, and Θ_ε is the $p \times p$ diagonal matrix of unique variances ε. In accord with matrix algebra, matrices are represented in factor analysis and structural equation modeling by uppercase Greek letters (e.g., Λ, Ψ, and Θ) and specific elements of these matrices are denoted by lowercase Greek letters (e.g., λ, ϕ, and ε).

The following equation reproduces the variance in the $Y1$ indicator:

$$\mathrm{VAR}(Y1) = \sigma_{11} = \lambda_{11}^2 \phi_{11} + \varepsilon_1 \tag{D.9}$$

where ϕ_{11} = the variance of the factor η_1 and ε_1 = the unique variance of $Y1$. For completely standardized model, that is, when variables are standardized, both ϕ_{11} and σ_{11} equal 1.00.

The model estimate of the covariance (correlation) of $Y1$ and $Y2$ can be obtained from the following equation:

$$\mathrm{COV}(Y1, Y2) = \sigma_{21} = \lambda_{11} \phi_{11} \lambda_{21} \tag{D.10}$$

For completely standardized solution, that is, when covariance would be interpreted as the factor model estimate of the sample correlation of $Y1$ and $Y2$, both ϕ_{11} and σ_{21} equal 1.00.

Estimation of CFA model parameters—the underlying principle of maximum-likelihood (ML) estimation is to find the model parameter estimates that maximize the probability of observing the available data if the data were collected from the same population again. That is, minimize the difference between the predicted (Σ) and observed (S) covariance matrices. If one or more of the factor indicators is categorical (or non-normality is extreme), normal theory ML should not be used because of its propensity to inflate χ^2 values and to underestimate standard errors. In this instance, estimators such as weighted least square, robust weighted least square, and unweighted least square are more appropriate. Currently, the Mplus program appears to provide the best options for CFA modeling. In this study, Mplus program was used with categorical data and the weighted least square parameter estimates using a diagonal weight matrix (W) and robust standard error and a mean- and variance-adjusted χ^2 test statistic (Brown 2006). In weighted least square parameters, the number of elements in the diagonal W equals the number of sample correlations in S, but this matrix is not inverted during estimation. Weighted least square estimation is fostered by N being larger than the number of rows in W.

Descriptive goodness-of-fit indices—The classic goodness-of-fit index is χ^2. Under typical ML model estimation, χ^2 is calculated as follows:

$$\chi^2 = F_{\mathrm{ML}}(N - 1) \tag{D.11}$$

In Mplus program, χ^2 is calculated by multiplying F_{ML} by N instead of $N - 1$. Thus, a statistically significant χ^2 supports the alternate hypothesis that S is different than Σ, meaning that the model estimates do not sufficiently reproduce the sample variances and covariances which were in this research.

Although χ^2 is steeped in the tradition of ML and SEM, it is rarely used in the applied research as a sole index of model fit even if is used as nested model comparison. Another index that falls in the absolute fit indices' category is the root mean square residual (RMR) which reflects the average discrepancy between observed and predicted covariances. However, RMR can be difficult to interpret because its value is affected by the metric of the input variables. This index was recommended by Mplus program because of categorical data.

Indices from the parsimony class (number of freely estimated parameters as expressed by model degrees of freedom—df) would thus favor model solution fit the sample data with fewer freely estimated parameters. A widely used and recommended index from this category is the root mean square error of approximation (RMSEA). The RMSEA is a population-based index that relies on the noncentral χ^2 distribution, which is the distribution of the fitting function (F_{ML}) when the fit of the model is not perfect. The RMSEA is then computed:

$$\mathrm{RMSEA} = \mathrm{SQRT}[d/\mathrm{df}] \tag{D.12}$$

where df is the model df (Mplus uses N instead of $N - 1$). The RMSEA compensates for the effect of the model complexity by conveying discrepancy in fit (d) per each df in the model. It is sensitive to the number of model parameters, but is relatively insensitive to sample size. RMSEA values of 0 indicate perfect fit (and values very close to 0 suggest a good model fit).

The noncentral χ^2 distribution can be used to obtain confidence intervals for RMSEA (a 90 % interval is typically used). The confidence interval indicates the precision of the RMSEA point estimate. However, researchers should be aware that the width of the interval is affected by the sample size and the number of freely estimated parameters in the model. According to Brown (2006: 84), specifically "close" fit (Cfit) is operationalized as RMSEA values less than or equal to 0.05. This test appears in the output of most software packages as the probability value that RMSEA is <0.05.

Comparative fit indices evaluate the fit of a user-specified solution in relation to a more restricted, nested baseline model. Typically, this baseline model is "null" or "independence" model in which the covariances among all input indicators are fixed to zero, although no such constraints are placed on the indicator variances. Some indices from this category have been found to be among the best behaved of the most of indices that have been introduced in the literature. One of these indices, the comparative fit index (CFI), is computed as follows:

$$\mathrm{CFI} = 1 - \max[(\chi_T^2 - \mathrm{df_T}), 0]/\max[(\chi_T^2 - \mathrm{df_T}), (c_B^2 \mathrm{df_B}), 0] \tag{D.13}$$

where χ_T^2 is the χ^2 value of the target model or the model under evaluation, $\mathrm{df_T}$ is the df of the target model, χ_B^2 is the χ^2 value of the baseline model or "null" model, and $\mathrm{df_B}$ is the df of the baseline model; max indicates to use the largest value whichever is larger. The χ^2_B and df_B of the null model are included as default output in most software programs. The CFI has a range of possible values of 0.0 to 1.0, with values closer to 1.0 implying a good model fit. Like the RMSEA, the CFI is based on the noncentrality parameter.

Another popular and generally well-behaved index falling under this category is the Tucker–Lewis index (TLI). The TLI has features that compensate for the effect of model complexity; that is, as does the RMSEA, the TLI includes a penalty function for adding freely estimated parameters that do not markedly improve the fit of the model. The TLI is calculated by the following formula:

$$\text{TLI} = [(\chi^2_B/df_B) - (\chi^2_T/df_T)]/[(\chi^2_B/df_B) - 1] \tag{D.14}$$

where, as with CFI, χ^2_T is the χ^2 value of the target model or the model under evaluation, df_T is the df of the target model, and χ^2_B is the χ^2 value of the baseline model or "null" model. Unlike the CFI, the TLI is non-normed, which means that its values can fall outside the range of 0.0 to 1.0. However, it is interpreted in accord with a good model fit.

References

Bartholomew, D. J. (1987). *Latent Variable Models and Factor Analysis*. London: Charles Griffin & Company LTD/New York: Oxford University Press.

Basilevsky, A. (1994). Statistical Factor Analysis and Related Methods. Theory and Applications. In W. A. Shewhart & S. S. Wilks (Eds.), *Probability and Mathematical Statistics*. New York: Wiley.

Brown, T. A. (2006). *Confirmatory factor analysis for applied research*. NY: The Guilford Press.

Gardner, R. C. (2003). *Estadística para Psicología usando SPSS para Windows*. México: Prentice Hall.

Jolliffe, I. T. (1986). Principal Component Analysis. In D. Brillinger, S. Fienberg, J. Gani, J. Hartigan & K. Krickeberg (Advisors) *Springer Series in Statistics*. New York: Springer.

Printed in the United States
By Bookmasters